HOW

TO

SMELT

YOUR

GOLD & SILVER

Revision 5

BY

Hank Chapman, Jr.

A Complete Plain English Guide
For The
Amateur Or Professional

DEDICATION

To my wife, for all those gentle nudges in the right direction.

First Edition Printed October, 1996
First Revision September, 1997
Fifth Revision May, 2019

Hank Chapman, Jr., 2795 Avenida Grande, Bullhead City, AZ 86442
E-mail hchap@suddenlink.net
Phone 928-758-2215

Published By:

Sylvanite Publishing
www.sylvanitepublishing.com

WARNING!!

FOR INFORMATIONAL PURPOSES ONLY!

Since the author and publisher of this book have no control over the way the methods and procedures outlined in this book are used, this information is presented for

Informational Purposes Only!

Unless the author is standing beside you as you use this information, you are on your own! Be Careful! You, and you alone are responsible for your actions.

The Publisher, Author and Retailer assume no liability whatsoever for any method in which this information may be utilized, or damages arising from the use of this information.

All chemicals, at a certain concentration, can be harmful to human life, or fatal. People die every year drinking a glass of water. Our bodies have to have water to live, yet water can easily kill you in the right circumstances. Read the information on personal protective equipment carefully, **and use the appropriate protective equipment.**

High temperatures, such as those used in smelting, will cause some metal oxides and chemical compounds to gas off into the atmosphere. These fumes (smoke) can suffocate and kill you, or other people. **Use adequate forced ventilation!** Do not assume that doing these procedures outside will create adequate ventilation. Assay fumes and smelting fumes are poisonous, and toxic to life. Again, **use adequate forced ventilation!**

Never mix chemicals unless you are absolutely sure how they will react. Never breathe dust from dry chemicals. Never breathe dust from silica. Never breathe acid fumes. Store any containers of wet chemicals or chemical solutions in a ventilated cabinet. All of these fumes and dusts will damage your lungs, and can cause silicosis or chemical pneumonia. Open or closed chemical containers will leak corrosive fumes that will damage electronic equipment, and cause corrosion.

Always assume smelting equipment is hot!

At smelting or assaying temperatures, the burns you can receive are instant third degree burns that will take months or years to heal. Keep pets, children and strangers away. Do your work in a secure area. **Never do chemical experiments, assaying or smelting in your home!** Never, ever smelt amalgam, or any material containing mercury. Never use or re-use any container that has had chemicals or chemical solutions in it. Never drink water or anything else out of a beaker. Maybe it had acid in it yesterday. **Use common sense and think about what you are doing!** Think each task through before you start, and use the appropriate safety gear.

Technical Support

Telephone support is available free for safety reasons, and is included in the price of this book.

Read the entire book before calling... the answer to your question is probably there, if you look for it! Have your question(s) written out before you call. It will save you time.

Support is available for smelting issues only. If you have process (milling) questions, read "How to Mill Your Gold & Silver" from cover to cover, then call, or e-mail to the following address:

Hank Chapman, Jr.
2795 Avenida Grande
Bullhead City, AZ 86442

Phone 928-758-2215
E-mail: hchap@suddenlink.net

Table Of Contents

Information for Revision 5

So, you ask, what's new in this book? There is information on induction furnaces and other equipment, and a focus on some analytical techniques. There is, in general, an update on the entire process of smelting. There are specific responses to feedback generated by readers throughout the book over the last ten years or so. And yes, less graphics, more pictures. Several readers have expressed an interest in shotting their metal, so a chapter has been added on that subject. A chapter on information has also been added to account for online sources of information.

This book will not cover any chemical process other than the very basics as related to the original premise of the book, which is to get your metal to the point that you can market said metal for a profit.

And, it seems, hopefully, with the election of Donald Trump as President, mining in general just might get a break from the ridiculous restrictions imposed for the last eight years. Governmental overreach became incredible to behold. All the attempts to push "climate change" and the "carbon tax" are beyond comprehension. This is truly a form of income redistribution, and the socialist agenda of our previous politicians.

And of course, the NOAA and others have been caught "cooking the books" on climate change. Science has always maintained that weather repeats in cycles. Such as 10 year storms, 50 year storms, and 100 year storms. Why do we not have 150 year or 200 year storms? For the simple reason that there are no records existing of weather much further back. The exception, of course, being the Biblical reference to a flood.

And then, we forget about one of the larger climate polluters. That would be Mother Nature, in the form of the volcano. And, of course, there's the Super volcano. Maybe the climate change crowd can figure out how to cork volcanoes. There are, in fact, 1500 potentially active volcanoes on land world-wide, and 50 to 75 or so that erupt annually. It is stated that there are about 20 or so erupting or active at any given time. It is actually unknown how many volcanoes exist. There are many thousands undiscovered on the ocean floors.

Granted, that does not mean that we should simply ignore pollution of any kind. Every country should do their part, just not expect America to pay for it. Without a doubt, the most significant factor of human pollution is simply excess population, too many people. And thus, the demand for goods and services. We, the human race, are breeding ourselves off of the planet, and not being good stewards of said planet.

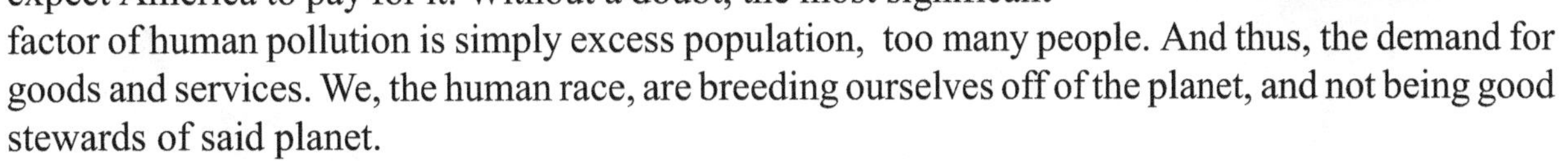

You should know that these days, the Agencies doing the monitoring don't even have to leave their desks. Pollution is monitored by satellite. A bloom on the screen, and they know where, and how much of whatever is being discharged. In fact, if you *apply* for a discharge permit in Nevada, they "put you on the satellite" and monitor your operation. It goes without saying that any and all discharges are monitored, permitted or not. Smile! You may already be on satellite.

Another amazing group is the environmentalists as a whole. They protest mining in their stone washed jeans, wearing gold wedding rings. They go to the protest in their Prius or Volt. Any metal in that Prius? Or the Volt? What they want, is for you to give up all your stuff, but they don't have to give up their stuff. They have copper wire in their houses, chrome plated brass plumbing fixtures and all those other mined products. Have you ever seen an environmentalist living in a cave, swathed in animal skins? That's what they want you to do, but not them, Heaven forbid. How many kids do they have? There's a whole lot of this hypocrisy being spoken, yet everyone is afraid to take them to task on it.

How much electricity does it take to charge that Prius or Volt? Ask, and watch the answers they come up with. What about those big lithium storage batteries? Where does lithium come from? Where do big lithium batteries go when they die? How do they think electricity is generated? And transmitted to the consumer? Gee, is that wire copper or aluminum? People really need to understand where we will be if these people run unchecked.

The people in industry are the people that develop better and better ways to keep any industry clean. And sooner or later, Mother Nature may decide to cull the herd. Who knows? Just remember that *if it isn't mined, it must be grown.* Ralph Waldo Emerson said "moderation in all things." We would all do well to remember that.

Good Luck with your project!

Smelting is about as old as history, dating back to King Solomon. Today, most mines, large and small, smelt various precipitates and other precious metal residues to bullion. Smelting, as referred to in this book, will refer to the process of treating precious metal concentrates, in one form or another, with dry chemical fluxes and high temperatures to collect the precious metals, and in some cases, upgrade, or refine the precious metal(s) in the smelt.

Traditionally, base metal ores were smelted to recover some base metals, such as copper or iron, by huge smelters all over the country. Today, steel mills still smelt iron ores after adding manganese, tungsten and other ingredients according to the grade of steel they wish to produce.

As things happen, the high temperature utilized in the smelting process can create a serious amount of pollutants that are can be discharged into the atmosphere. Sulfides gas off as sulfur dioxide, SO_2, carbonates create carbon dioxide, CO_2, and so forth. The primary sources of SO_2 and CO_2 are not smelters, or smelting operations. Look at the automobile, diesel engines, and other chemical processes. Fossil fuels are another major source of pollution.

As time went by, and environmental restrictions began to emerge, high volume smelting, by and large, with the exception of steel mills, has pretty much fallen by the wayside in this country. The lower grades of ore being smelted produced a large amount of impurities such as arsenic oxides, SO_2, and CO_2 that were discharged into the atmosphere in violation of the smelter's discharge permit. With higher grade ores, the smelter operator simply passed the cost of the fines for violating the discharge permit to the mine. As the years went by, tighter and tighter restrictions were placed on the smelters, with higher and higher permitting costs, until the smelter was driven out of business. Today, most smelting of an industrial nature is done in Canada, or Mexico. Environmental regulations are still lax enough that smelting is permitted in these two countries on this continent.

Since the world-wide environmental movement is rapidly growing, and the discharge of SO_2 into the atmosphere is largely responsible for the continuing creation and spread of acid rain, it is but a matter of time before the large smelters are a thing of the past. Still, it is hard to believe that by crossing an international border, an imaginary line only inches across, you can smelt most anything, spewing gasses into the air on both sides of the border. Smelting is still common in most countries. What has been accomplished by regulating one country while the others are trying to catch up, and smelt more and more every day?

As common as this process is today, it is difficult to find any literature on the subject. Thirty or forty years ago, smelting literature was common. Finding an old text in a used book store is about the only hope. A lot of competent assayers have figured out methods that work for them, and soon move from the assay lab to the refinery. Most mining companies are not real eager to discuss the methods they use, or the fluxes, or even the temperatures they smelt at. Some mines fear theft, which does happen, and do not want the public, or anyone else to know that radioactive tracers are added to their bullion during the smelting process. This addition makes the recovery of the precious metal more likely in the event of theft. Word does have a way of getting out, and those serious students of mining and prospecting will find a way.

Strangely enough, most of the gold mines in Nevada are using pretty much the same flux as what is outlined in this publication. For the most part, the silver mines are using minor variations of the silver flux. None of the mines really try to ship a super high-grade bullion to the refiners. It is seen as a matter of efficiency, and considered a waste of time to try to refine much past .800 fine. After all, that's what the refiner is for. Each mine has an arrangement with the refiner as far as penalties and refining costs go. Typically, four percent (or more) of the gold is charged off as a refining fee. Sometimes more, sometimes less, depending on volume, purity, and other parameters. Mines will not discuss the arrangement they have with their refiner at all. Strict secrecy prevails. If you think you can call a refiner on the telephone and get any information at all, with the exception of an appointment, you're in for a surprise. Plan on meeting them at their place of business, with a sample of what you produce, where you will discuss terms. Your terms will probably be a lot different than a large mining company's terms.

Don't waste time by asking exactly how they refine the metal. You might hear a few real generic terms, or hear about obsolete refining processes, but no refiner in their right mind will disclose procedures. These are closely held, proprietary techniques.

What we intend to accomplish with this document is to provide enough information for you to smelt, or "fire polish" your gold or silver to the point that you will not necessarily need a refiner to buy your product. This can open up new markets to the small or mid-sized mining company not currently being utilized. Selling your product is probably the easiest thing in the world to do once you understand how all this is done. You must be completely honest, and produce a very high quality product to do this, however. The first time you misrepresent the fineness of your metal, knowingly or otherwise, you will become a pariah in the mining industry, shunned by all. Better to play it straight and be proud of what you do, and your craftsmanship.

Remember that gold and silver are simply commodities. The IRS considers precious metals commodities, and assets. So should you. If you get "gold fever" you will also get more trouble than you could possibly want. And it does happen to a lot of normally sane people. One look, and they get totally stupid, and have even been known to kill people. Don't let this happen to you. Watch the movie "Mother Lode" with Charleton Heston. That movie is fiction, and actually pretty tame compared to some of the stupidity that has went on in the last ten years or so. Some of these war stories are true. Really nasty things do happen. Gold can bring out the very best or the very worst in some people. Hopefully, you will be the former.

Whatever you do, pay close attention to the chapter on safety. Very few people are injured doing what is described in this document. Typically, the people who smelt and refine their metals are very serious about the methods, and are very meticulous in practice. This is not a process that can be "slopped" around. It's like the computer saying, "garbage in, garbage out." Poor workmanship will produce poor results. Poor safety habits produce serious injuries. Get in the habit of using protective gear at the onset, and you will go far. Sooner or later, a "pro" will come along, and you will think you have a kindred spirit at hand, until he laughs at your safety gear. Be patient, sooner or later, you'll see his burn scars. So who is the pro? Just remember that this is an industrial process, with inherent hazards you must minimize. Be safe in your work. When in doubt, ask first, and act later, if that's what it takes.

Good luck!

Chapter One

Information

Online And YouTube

When this book was originally written, the Internet barely existed, and YouTube was just a twinkle in someone's eye. Now, the Internet and YouTube are the "go to" places for information. The "good old days" of cracking a book seem to be gone forever. Instead, watching a YouTube video is considered an education on any subject, including smelting.

Granted, there are some excellent videos on YouTube and a wonderful source of information for things such as fishing knots, and other general subjects. There are lots of videos on smelting and refining that are simply mind boggling when one considers that there are some of these videos that are just wrong, and can lead to some horrible consequences. An interesting video is out there on the Parks Method of processing galena, a lead-silver sulfide. The author of the film would have you dig a hole in the back yard, layer the hole with charcoal briquets and pans of galena over the briquets. A pipe is used to allow a blower (output side of shop vacuum) to blow air to the briquets, and in essence, create a smelter.

The author instructs the user in a not so authentic hillbilly drawl through the steps, but never mentions that you are contaminating your back yard with lead forever and ever. He also never mentions that the fumes are going to give you lead poisoning, assuming lead is the only metal that will oxidize to vapor. All of this contamination to recover a few ounces of silver. You will have, however, created your very own, very expensive to clean up mini superfund site.

A website that used to be on the Internet that taught people how to smelt gold lasted about six months. Your author was appalled at some the practices used and taught with a complete lack of safety. No ventilation, and the instructor was smelting concentrates from ore. Sure enough, the last post was an obituary for the trainer, announcing that he expired from thallium poisoning.

The point here is that you get what you pay for, and there is no substitute for a little training and self education. You may put the video on your smart phone, and you can watch it ten thousand times, but at the end of the day, if the subject matter is incomplete or inaccurate, your process will be just as flawed as the video. Any injury or chemical contamination is on you.

As you read through this text, there will be information provided for you about ascertaining what is in the material you are smelting. The first order of business is you to know and understand exactly what you are dealing with. Do you have any idea what you have? Is it radioactive? Does it have thallium, osmium, beryllium or lead in it? Granted, once you know what you have, you can use the web, and Google the element in question for more information. But how do you find out what you have? That information is in this book, but never on YouTube will you see or hear the phrase "multi element spectrographic analysis" or any indication that you need this analysis.

Information is what keeps you safe, so get all the information you can. For something as complex as smelting, YouTube is mostly incomplete. The Internet has a huge amount of information on smelting. How much are you willing to watch to keep yourself safe?

The expression we are after is "*Caveat Emptor*", let the buyer beware. You should at least learn enough of the subject matter at hand to determine if the information is good or bad.

Books Online

Here are a few web sites if you are looking for a book or two. And it never hurts to have a hard copy of the information you are referring to. If you choose to have a book for reference, you can check the following web sites for books that interest you.

Miner's Inc: www.minerox.com
Mining Books: www.miningbooks.com
Make Your Own Gold Bars: www.makeyourowngoldbars.com
Legend, Inc: www.legend-reno.com
Action Mining Catalog: www.actionmining.com

Recommended Books: A Selected List

Here are some of the books you might want to have. The primary focus here is on the reference materials for you that will help your operational problems. Review the description of the book, it may help you with information you need. If you are interested, try the online vendors above, or Google the title. Obviously, if you don't think you will need the book, don't buy it.

Standard Methods of Chemical Analysis, N. H. Furman, Editor

There are five volumes in this set, however the first volume will cover everything the aspiring chemist or assayer would want dealing with chemical analysis of minerals. There is also a very good section on the Fire Assay. There are analytical techniques for all the elements here. This book is probably out of print, however there are some on the web as government surplus.

The Metallurgy of Gold, and The Metallurgy of Silver, by Sir T. K. Rose

Anything and everything about gold and silver can be found in these two books. The information is quite extensive, and if you think the old timers didn't know what they were doing, well, these books will change your mind. The books cover all aspects of gold and silver processing and equipment through the 1930's. You will notice a lot of the old equipment was taken out of these books, and renamed, somewhat modified, and is in use today.

Fire Assaying, by Shepard and Dietrich

This book has been reprinted, and is available at most of the web sites previously listed. If you are at all interested in the fire assay, this book should be your first choice. A lot of people buy Bugbee's "A Textbook Of Fire Assaying", and feel like they are trying to decipher Greek. Read Shepard and Dietrich first, then try Bugbee. Since assaying is close to smelting, an assay manual can be useful. Just remember, no lead in your smelts! Never create a separation problem for yourself.

Recovery And Refining of Precious Metals by C. W. Ammen

This is an excellent reference book. There are a lot of chemical procedures, chemical tests, and information dealing with lab techniques. Apparently this book has been recently revised, so make sure you get the latest edition. It is easy to understand, and explains a lot of general chemical terms, as well. There are lots of illustrations, a good glossary and a decent amount of information on assaying.

Anatomy of a Mine from Prospect to Production by the US Department of Agriculture (Forest Service)

General Technical Report INT-GTR-35. If you are new to mining, or contemplating an operation, especially on Federal land, this is where you should start. There are sections covering Mining Law, Prospecting, Exploration, Development, Production, and Reclamation. All from a government land manager's perspective. It is downloadable from the web, and is discussed in further detail in this book. Go to www.fs.fed.us/rm/pubs_int/int_gtr035.html or Amazon.com. Amazon will charge for the report.

SME Mineral Processing Handbook, Volume 1 & 2, Edited by Weiss

These books pretty much replace the "Handbook of Mineral Dressing" by Taggart, so don't despair if you can't find Taggart. These books cover as much as Taggart, and are more current. They are published by the Society of Mining Engineers. Lots of useful information on grinding ores, process and such topics.

Condensed Chemical Dictionary by Hawley

This book is handy when you need to look up a chemical, and find out more about it. Also has the chemical formulas for all the chemicals. If you are going to leach with any chemical, you'll need this one. Dry reagents used in smelting fluxes are also listed here.

Chemical Technician's Ready Reference Handbook by Shugar & Ballinger

This book takes you through basic lab procedures, glassware, safety, lab math, and on and on. How to be a lab technician, and what to do, pretty much. Do you know how to fold a filter paper? It's in here. If you're doing any wet chemical work, you need this book.

Introduction To Chemical Principles by Stoker

This book is basically a Chemistry 101 text, and this one is not an absolute necessity. Your author has three of these "introduction" type chemistry books, two are the 101 type, the other is a 202 type book. The idea is to familiarize yourself with some chemical procedures so that you can safely work with chemicals. Understand what you are dealing with. You will also find Avogadro's number in these books. There is also a section on Pyrometallurgy.

Analytical Chemistry For Technicians by John Kenkel

This book is a must if you are interested in analytical chemistry. The book covers all the methods and equipment used in analytical work, as well as maintaining a notebook, titration, and most anything analytical. A lot of good information on the elements, as well.

Local Publications

Keep in mind that each state has books on mining. For example, you are interested in a mine or mining district in Nevada. Try this one: "Mining Districts and Mineral Resources of Nevada" by Francis Church Lincoln. This one even has a map, as well as all the historical data on the old mines and what mineral was mined.

There are also books about sections of states and counties, such as "Mines of Eastern Nevada", and "Mines of Storey County". These types of books can narrow a search on a district or mine, and will typically provide more detail than books about statewide mines.

A reference library can be useful.

Chapter Two

Safety

General:

This is where it's at. Pay attention here, and you will go far. Invest in the safety equipment described below, as much as you can afford. The equipment is totally useless without common sense to go with it. Some of these items can make your work much more pleasant, as well as safer. A good example is the gold-film covered face shield commonly used in fire assay labs. The heat of a furnace at smelting temperatures can have quite an impact on your face when you open the furnace door, or begin the pour with a tilting furnace. With the face shield, the gold film reflects the heat, and shades the white-hot glow from the interior of the furnace. This way, you can look into the furnace, or into the crucible with no problem. We will cover all the aspects of the equipment you will need, and point out other areas you should be thinking about, as well. Have all your safety equipment cleaned, checked, and available before you start. Re-read this chapter a time or two as you go, become totally familiar with the contents.

Be safe! Live to spend it!

Protective Clothing:

Race car drivers wear Nomex clothing in case they crash and burn. You should wear similar protective clothing when working with metals at high temperatures. Most safety companies carry Nomex, or similar heat resistant garments from jackets to pants to booties to cover your feet. There are even sleeves made of heat resistant material that you can pull on over your shirt. The bare minimum is a jacket. This gear is warm to start with, and being around high temperature furnaces makes it a lot warmer. Hot or not, wear at least the jacket. Never, ever wear Nylon, Dacron, Orlon, Polyester or other synthetic fiber clothing while smelting. The heat will melt the cloth right into your skin. Most assay labs have at least one scarred veteran that has welded Nylon to his or her shoulder. Fire assayers spend a fair amount of time in front of a furnace with the door open, and if they don't wear protective, heat-resistant garments, the shoulder nearest the open door starts smoking about the time they finish pouring the set they are working on. Normally, you will smelt at higher temperatures than those used for a fire assay. So, at least a jacket will start you on the right path.

Nice, but not what we had in mind.

If you ever drop a crucible full of molten, super-heated metal and flux, you will quickly come to appreciate the heat resistant pants, jacket, and booties, as well as eye protection and face shields.

It is spectacular to watch the little fires start everywhere the metal and slag hit. That includes the human body. It is very impressive to watch how fast people can undress under these circumstances.

Never wear lace-up shoes if you can avoid it. Wear pull-on steel toe boots, and keep the pants cuff outside the boot. Steel toe boots are a must working around heavy objects that fall on your feet. Pull-on boots are handy when white-hot slag spills onto your foot. Think about unlacing a boot while your foot is cooking. Pull-on boots, even if they have pointy toes, are a must. If you insist on lace-up boots or shoes, make sure they are high top, or you have the heat resistant booties on. Hot, sharp slag will jump into low top shoes.

The latest in Steel Toe Boots! (Spurs are optional!)

Welder's clothing can be useful, as well. It is easy to find, and designed for working around high temperatures. Look in the Yellow Pages under "Welding Equipment and Supplies".

Gloves:

Assay supply houses sell shoulder length and elbow length gloves and mittens that will withstand 2000° F. Forget the gloves, the material is so heavy your fingers won't bend in them anyway. The mittens are great. The shoulder length mittens are the best. There are usually two grades of material, with the better grade being somewhat more expensive, but well worth the expense. Since your hands will be the closer to the heat than any other part of your body, make sure they are adequately protected. Use two mittens, one on each hand. The first time you use only one, you will understand why two are necessary.

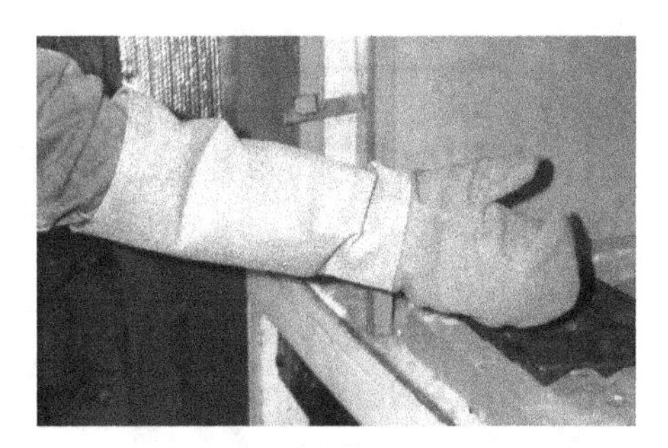

Elbow length high temperature mitten.

Always keep a pair of good, heavy leather gloves handy. They are great to have when you're handling hot objects, or swinging a hammer to break some slag loose. Always wear gloves when working with hot or cold slag. Remember that slag, as part of the cooling process, will spall (as in explode) violently, sometimes several hours after cooling. Some slags will spall again as atmospheric conditions change. Sweep slags up as soon as possible, and keep them in a covered metal container until you are ready to dispose of them according to local, State, and Federal regulations. The slag you create is basically a borosilicate glass, and is so sharp it will cut you with no pain. You become aware you are cut when you bleed. Have a healthy respect for slags, and never hold slag near your face to look at because the colors are pretty. Keep those gloves on. Use forceps or tweezers if you must handle slags. Read about slags containing beryllium later.

The latex rubber gloves used by physicians are a must to have on hand when weighing up chemicals such as flux ingredients. Do not buy sterile gloves. Vinyl examination gloves are more than adequate for our purposes. These gloves are cheap, and available at most safety outlets, by mail order, or at medical supply houses. Some flux ingredients have a strong pH, and contact with the skin can cause skin irritation, or with some ingredients, skin burns. At the very least, skin irritation is likely. Check the gloves for pin holes, tears and leaks before you put them on. Blow the glove up with air. If it doesn't hold air, don't use it.

Eye Protection:

If you spend any time at all around fire assay labs, you will encounter some macho fool wandering around with slivers of slag stuck in their face. They pour a set, and slag 'em down with no face shield, usually just safety glasses. They aren't even aware the slivers are stuck in their face, until someone laughs at them, or asks if they can feel the slivers. If those slivers of slag have any appreciable amount of beryllium in them, a tumor will grow where each sliver of slag stuck. Got beryllium?

Safety glasses are a necessity. Wear contacts? Take them out when you are working around the fumes and heat. The heat can instantly dry them out, and dry chemicals, such as flux dust can get behind the lens and do serious damage to your eye before you can get the contact lens out. Never, ever wear contacts when you are mixing chemicals, wet or dry, or smelting. Or when you are assaying, for that matter. A pair of cheap prescription glasses is much better than contacts, and safety glasses are the best bet. You were only issued two eyes at birth. Don't take a chance with them because of vanity, or whatever. If you think you look like a dork wearing glasses, well, just don't invite anyone in to watch while you work. Protect your eyesight. Don't wear plastic sunglasses, they will melt and run down around your ears. Use the gold film-covered face shield over your safety glasses. You are smelting in a safe manner, not trying to be cool. Forget the sunglasses. The gold film-covered face shield (Oberon)can be hard to find. A local assay lab supply house has them. Look for Legend, Inc. in the supplier's Appendix at the back of the book. They ship anywhere.

Working around a hot furnace all day can still dry your eyes out. A small bottle of artificial tears from the local drugstore can relieve a lot of discomfort if you have this problem.

Respirators:

Respirators are incredibly uncomfortable in a hot environment, and can take a lot of getting used to. Now you know why a lot of fire assayers have such a high blood/lead content, and a lot of people smelting metals have tumors develop. A respirator with a set of new dust cartridges is the only defense against the airborne metallic oxides, such as lead. If you are smelting material with mercury in it, get yourself a set of mercury cartridges for the respirator. Always match the cartridge to the job at hand. Most importantly, remember that ***respirators do not supply air.*** A Self Contained Breathing Apparatus (SCBA) supplies breathable air. Respirators filter the air through the cartridges, thus eliminating dust, or certain vapors. Certain metallic oxides are very, very toxic. Think about Thallium. This was used as the main ingredient in rat poison for years. It is so toxic that the poison manufacturers were forced to use other chemistry. Guess what? Thallium is

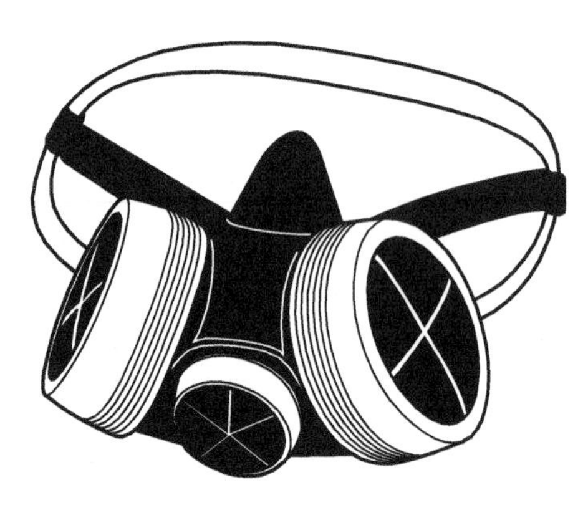

Wear your respirator!

common in nature. Any massive deposit of ore in this country will most likely have substantial amounts of Thallium included in the deposit. Also be aware of Cadmium. It is incredibly toxic, and like Thallium, fairly common in ores. There will be other dangerous contaminants listed further on.

Smelting fumes can kill!
Use adequate forced ventilation.

All smelting fumes or assay fumes are highly toxic, and poisonous. As the intense heat breaks down various compounds, such as sulfides, the metallic oxides will gas off. Any material that has been chemically pre-treated will also create toxic fumes. Osmium is a killer, lead will seriously damage your body, and mercury vapors are very toxic. Lead poisoning causes brain-damaged babies, and causes problems with the reproductive process. Did you macho types catch that last sentence? A few people die every year from retorting amalgam on the kitchen stove. Remember that any process involving any metal at high temperature will evolve toxic fumes. Mercury *starts* vaporizing at room temperature, and should be stored in an unbreakable sealed container, under water or oil. If you work with mercury, never expose you r bare skin to it, and wear your respirator with mercury cartridges. Make sure you have adequate forced ventilation in the area where you are working.

A lot of geniuses smelt outdoors (no respirator, of course!) since the breeze will blow the fumes away. Away from where? Doesn't the wind still shift? Or maybe the wind only blows in one direction there...What happens when the wind shifts and blows your house full of fumes? Or blows the fumes into the intake on your air conditioner?

Protect yourself from a short agonizing death, or a long, agonizing slow death. Wear your respirator if you even think of getting near a smelting or assaying operation. A good rule of thumb is to know what elements are in the material you are smelting. A simple spectrographic analysis will tell you what is present in your material, and could save your life. A simple water scrubber on your ventilation system is a good idea, as well.

Ventilation:

Ventilation was touched upon in the previous section, however, there's more. Positive forced ventilation is what we are after here. Visit a chemical lab if there are no assay labs in your area. You will notice that the fume hoods pull any chemical vapors toward the back of the hood, away from the operator. There will be two or more slots in the back of the fume hood, one or two at the bottom, and one or two at the top. The slots at the bottom pull the fumes back and away from the operator, the slots at the top pull any fumes that get past the bottom slot out of the hood.

The hoods over furnaces are usually suspended fairly close to the top of the furnace, and are larger than the furnace by a foot or so on all sides. There are usually no baffles or anything of that nature inside the hood. It is a hollow shell that simply collects the fumes and moves them away from the furnace, and thus, the operator. A large blower usually is mounted above the hood, either on the roof or in the attic, and the outlet usually goes to a bag house, or just vents through the roof to the atmosphere. Typically, smelting fumes are corrosive, and a fortune can be spent on stainless or corrosion resistant blowers and ducting. It is a lot cheaper for the small operator to use plain steel or galvanized ducting and plan to replace every few years, depending on the usage of the equipment. Blowers are usually painted with epoxy paint, or a similar finish, and will last quite a long time. If you are repairing or replacing your ducting, and find dust adhering to the inside, don't throw it away until you have the dust assayed, or assay it yourself. Flue dust, as it is called, can contain a surprisingly large amount of values, as in precious metals. You might be smelting too hot if your flue dust has high values. Always mix your flux under your fume hood, while wearing your respirator. The dust from the flux ingredients will irritate your lungs, if you breathe it.

You can check the operation of your ventilation system by using the commercial smoke bombs available at most safety supply houses. Just follow the directions on the package. If you live in a colder climate, you might want to vent fresh outside air in near the smelter to prevent heat loss from your regular room heat. You can also vent fresh air in anywhere you want it, if you live in a warmer or hot climate. Get a W. W. Grainger's catalog and look at the blowers they offer. Everyone knows someone who has an account with Grainger's. Grainger's has supposedly lowered their sometimes high prices on their blowers, which are excellent. This is great for the do-it-yourself types.

Smelting fumes can kill!
Use adequate forced ventilation.

The type of blower used will depend a lot on the type of power available in your building. Multiple phase motors, such as 220 three phase, consume about half the amps of 220 single phase. Blowers using 110 volts will be too small, in most cases, and are the most expensive to operate. Use a blower that will compress the air, such as a squirrel cage blower, or a paddle-type blower. These blowers move a lot of air, and can overcome high static pressures in the ducting, making them ideal for high cubic feet ventilation systems. Try to change the air in your smelting room at least once every three minutes. The fumes from your furnace should go directly out of the hood. If the fumes are collecting in the hood, you need to move more cubic feet. Calculate the cubic feet in the room, length times width times height, in feet, and there you have the number you need, the total cubic feet you must change at least three times a minute.

Don't just move air, exhaust it to the outside.

You can hire a professional to come in and do the work. Some of these guys are very good, and not too expensive. Some are very expensive, and not very good. Get estimates if you go this route. Some of the simplest ventilation systems in small rooms are swamp coolers that have been reversed to pull air out, instead of in. Simple, but effective, since swamp coolers are cheap CFM's, and use squirrel cage blowers. When it comes to ventilation, it is better to err on the safe side. If for a few dollars more you can get a slightly larger blower, do it. Another way is to run a multiple speed blower, such as a furnace blower. These are handy. You can run at a lower speed until you're ready to pour, then run at a higher speed when pouring. You generally will see a lot more smoke, or vapor when you pour since you are exposing the molten metal to air.

If you are renting a building, talk to your land lord before you cut a series of ten inch holes in the roof. Some landlords get real excited if you carve the place up, since they figure you'll forget to repair the holes in the roof when and if, you move. This must be where the phrase "damage deposit" came from. Older buildings usually have high, small windows that are handy to vent through. The only drawback is that a sheet metal shop will have to fabricate a square to round adapter for you, or you will have to place sheet metal in the opening and run the ducting through the sheet metal.

The temperatures inside the duct are low. Enough free air is available to keep the heated air around the furnace cool, so single wall ducting is usually adequate. You should be able to touch the ducting without burning yourself. You can check with the Fire Marshal in your area, but usually these guys freak out when they find out what temperatures you are using. They think you will be putting 2100° F temperatures directly into the duct work. Make sure they understand what you are talking about if you talk to them. If they are the usual pompous bureaucrats so common these days, plan on problems and delays in starting up your operation.

Fire Hazards:

Hot crucibles, hot slags, and hot metals will ignite anything they come in contact with, including your flesh. Head for the brickyard, or the building supply house for some firebrick. Ideally, your furnace should be sitting on a concrete floor, or on a steel bench. Make sure your pouring mold is sitting on firebrick, metal, or concrete. If your slags overflow the mold, and this is common, be sure that whatever contains the slags is fireproof. Set your crucible on a layer of firebrick after you pour. Never set the crucible on a steel surface, since it will act as a heat sink and the whole steel structure will heat up. This is when you will get burned. Understand the temperatures you are working with, and act accordingly. Make sure that hot, smoking crucible is under the fume hood after you have poured. It will smoke for five or ten minutes, so be prepared to ventilate the fumes. If you are in a rented building, and your excess slag runs on the concrete floor, thermal shock will cause the concrete to break down and begin to disintegrate. Your landlord will not be a happy guy when he sees the damage, so catch the slag before it runs onto the floor in a heavy metal container. A mold inside a shallow one quarter inch thick steel tray will work nicely. Or a small mold inside a larger mold will work. Think the design through when you set it up.

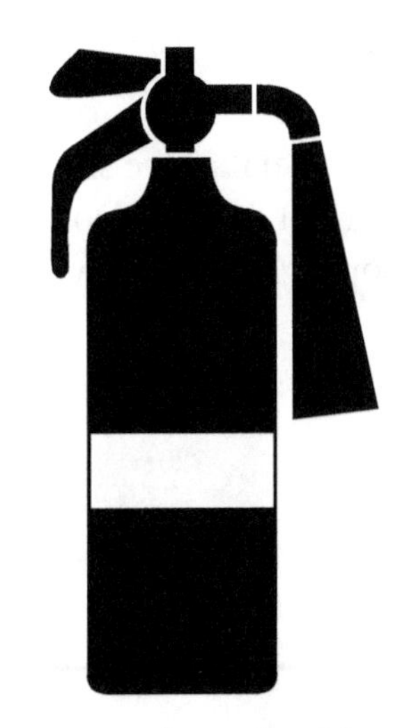

Make sure your fire extinguishers are accessible at all times.

The concrete board placed around wood stoves as an insulator can be useful for floor or wall protection. Concrete board doesn't burn, and that's what we are after. This product is carried at most building supply yards and wood stove dealers.

Burn Hazards:

Assume everything is hot. This will save you some time and pain on down the road. You can't imagine how many people walk into an assay lab and pick up a hot crucible or cupel. It seems you have to do this every year or two, just to remind yourself how bad those high temperature burns hurt, and how long they take to heal. So, assume everything is hot.

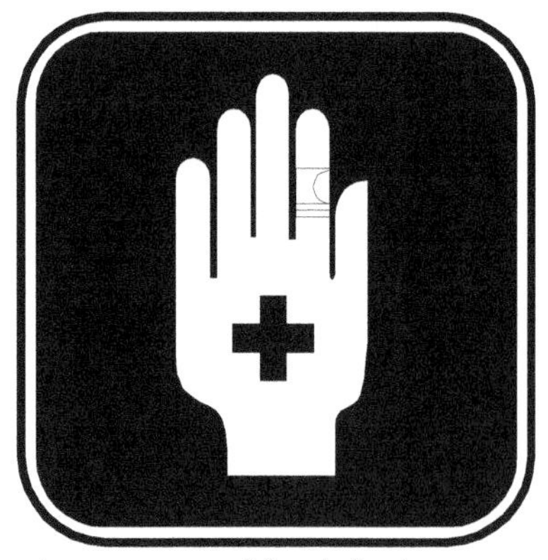

Assume everything is hot!

Invariably, when you cast or pour some metal, especially gold, someone will pick it up, and burn the hell out of themselves. Read this carefully! Smelted metal is superheated, and can burn hours after it has been poured. Never, ever handle the ingot, or metal, ***until you, personally, have quenched it in cooled water!*** Allow at least and hour for the metal to cool before you handle it. Quench it repeatedly in cold water. Then quench again, to make sure.

Never pour to a wet or damp mold! The slag and superheated metal will explode violently out of the mold from the formation of steam under the pour. Explode is a very accurate description of what happens. Your metal will be blown everywhere, never to be totally recovered. You will have instant fires wherever the molten metal or the slag land on something flammable. *Preheat the mold*. Set it on the furnace, or another hot surface. Don't put it in the furnace, you're going to have to handle it. We just want to dry it thoroughly before we use it. Use common sense. Hot things burn your tender body.

A large slag pot.

Chapter Three

Chemicals And Reagents

General:

Chemicals can be really nasty, wet or dry. People have a tendency to consider dry chemicals "safer". Well, they aren't, so don't get caught. Dry chemicals can stick to skin moistened by perspiration and burn the skin. The dust from dry chemicals can cause chemical pneumonia, and damage the lungs. So, consider all chemicals hazardous, at least until you know what you are dealing with. Wet chemicals are just as bad, they splash on you or your clothes, and you've got a problem. The vapors from either wet or dry chemicals can do serious damage, so think seriously about ventilation.

Fortunately for us, there are Material Safety Data Sheets (MSDS). When you order the chemicals you need, they should come with an MSDS. If they do not, call and request one from the supplier. According to law, all chemicals are to be accompanied by an MSDS, and the reason why will become obvious the first time you read one. The MSDS will specify all the hazards involved with the chemical, and tell you what protective gear is required to work with the chemical, among other things. **Read your MSDS very closely when you get it.** A lot of effort went into the program, mostly by worker's right-to-know groups. The MSDS will also tell you what to do in the event of a spill, and in most cases, how to dispose of the spent chemical, or chemical waste. So they are very handy, indeed. Don't throw the MSDS away when you have read it. File it for future reference, it contains valuable information that can save you or someone else a lot of discomfort, or even death. Read Appendix A to learn more about MSDS's.

Shortly, you will see a list of chemicals that you will need for smelting. Pay attention to the terminology, and try to understand some of the basic procedures. Observe all the safety precautions. If you can't understand what this is about, call the numbers in the front of this book, and have it explained to you. We all seem to know someone who is involved with chemistry in one form or another, so seek out that person and ask questions. **If you do not understand what you are doing, or what you are reading, don't do it!** Check the glossary at the back of the book. Go to a local high school and buy a chemistry text. Your best defense against injury is to educate yourself.

Don't be afraid to hit the books!

Always store chemicals properly, and according to local codes. If you're not sure, check with the local fire department. There are also excellent books on chemical storage available from chemical supply house.

You will note the word "anhydrous" when referring to the chemicals list. This means "without water". Here's the official definition: *(of a chemical compound) with all water removed, especially water of crystallization.* In other words, dry. Bear in mind what seems dry to you may still have a certain amount of water present, especially at the molecular level. When you buy anhydrous chemicals, they will come in sealed containers. Take what you need, and reseal the container. Some dry chemicals will actually pull moisture out of the air to the point the chemical will actually start to run water. The chemical is *hygroscopic*, and should be kept in a tightly closed container. This will also prevent contamination. Here's the official definition: *absorbing or attracting moisture from the air.*

The reason we are interested in the two definitions above is because moisture is what we don't want when we smelt. Extremely low levels of moisture will not affect the smelt, however any moisture of any consequence will cause the smelt to boil in the crucible. If there is enough moisture, the major portion of the contents of the crucible will be on the furnace floor, or on the floor, period. So, let's use anhydrous chemicals for our fluxes, keep all our chemicals and fluxes sealed in airtight containers, and make sure anything (such as molds, tongs or other equipment) we use is dry as well.

Chemicals also come in various grades. The very best grade is "USP", or Medical grade. USP chemicals are used in the manufacture of pharmaceutical products. Very expensive. The next best grade is "Reagent Grade", with very low impurities, and the next highest price. The next grade is "Technical Grade", which is what we are after. Some impurities, but not enough to really affect what we're going to do, and priced considerably l lower than reagent grade, or USP. The next lower grades are generally considered unacceptable due to the amount of impurities present. Remember "Technical Grade" when ordering your chemicals. Label your chemicals. If you take a small quantity out to use, label the container you put the chemical in. You will notice the words "white crystalline powder" in the paragraphs that follow. Two days after you mix a batch of flux, you will have small containers of white crystalline powder everywhere. If you don't know what it is, destroy it. Label those containers. There should be no doubt what is in any container at any time.

Common Smelting Reagents

Sodium Carbonate, Na_2CO_3 (anhydrous) (Technical Grade) is also known as soda ash, or washing soda. White, odorless, crystals or crystalline powder. Low in toxicity, used in water treatment, photography, pH control of water, glass manufacture, and bleaching of textiles, among other things. Technical grade is fine for our purposes. The dust will irritate the lungs. Note that some people think washing soda is an acceptable substitute for sodium carbonate. Not true. A lot of the washing soda on the market and available has ingredients added to make it flow better, and are not packaged to exclude moisture. Use sodium carbonate. Don't substitute. Sodium carbonate is an alkali, or basic flux ingredient. It is a primary ingredient for creating the flux. Read the MSDS that comes with it, and store accordingly.

Check the MSDS!

Silica, SiO_2, Silicon Dioxide, is also called silica sand. Silica sand is used in everything you can think of in one form or another, and is very common, and cheap. Your local building supply has it, hopefully in the right size. We want the silica reasonably fine, say at least minus forty mesh (-40). Ideally, some at minus one hundred mesh mixed half and half with minus forty mesh allows us to "stage" the smelt. Silica is a strong acid flux ingredient. The fine silica melts fairly quickly in the smelt, and the coarser silica melts later at higher sustained temperatures, giving a more uniform smelt. Normally, the silica used in assay fluxes is run as a blank to determine the gold content. We don't care if the silica has gold in it, do we? A little extra gold won't hurt our smelt. Normally, the gold present (if any) is so low in value it is of no consequence. If you order in finely ground silica, it can be fairly expensive. You can pulverize your own if you have the equipment, but beware the dust. Exposure causes silicosis. Silica sand usually does not have enough moisture in it to cause problems. If it does, dry it at 250°F until a glass object placed on the sand does not show steam, and you're on your way. Silica is a primary ingredient for creating the slag, and slagging iron based minerals. Store in a dry place.

Borax Glass (Anhydrous Borax) $Na_2B_4O_7$ (Glass), is made by calcining borax, a natural hydrated sodium borate found in salt lakes and alkali soils. We want to use the borax glass in our smelt, since it contains no moisture. This is the same stuff the famous mule team used to haul out of the desert before it is calcined. It is an acidic flux at the temperatures we will use. Regular borax off the shelf at the grocery store will work, but it will boil due to the moisture content. Borax glass works much better, and isn't nearly as messy to work with. Borax glass is a very common ingredient in assay fluxes. Check your local assay supply house. Borax Glass is a primary ingredient for creating the slag. Store in a dry place.

Manganese Dioxide, MnO_2, (Technical grade), also called manganese peroxide, and manganese black, is the most expensive of the flux ingredients we will use. It is a very dense black powder, and is derived from the ore pyrolusite. It is a strong oxidizing agent, and can ignite organic materials. It can be explosive in the right circumstances, so pay attention! Keep it tightly sealed in the original factory container. It is used in pyrotechnics, glass manufacture, textile dyeing, match manufacture and other things. It will color glass from a light purple to a purple so dark it will appear black, as you will see. We are interested because it is a strong oxidizer, and will take the impurities from our melt and put them in the slag. Read the MSDS and store accordingly. Your local ceramic supply house will have this. Go for a ceramic supply wholesaler, and you will save a buck or three.

Dry chemicals can be dangerous! Check the MSDS.

Calcium Oxide, CaO, lime, or "quicklime", as in Type S lime (unslaked) available at your building supply yard. A fine, white powder derived from limestone by calcining. A mildly alkaline (basic) flux ingredient, which we will use as a thinning agent in our flux. The dust will irritate your nasal passages and lungs, so wear a respirator when working with lime. Lime is used as a flux in steel manufacture, and used to neutralize acids. Lime was used for stage lighting before the advent of electricity, hence the word "limelight" and the phrase "in the limelight". Lime is not a primary ingredient in the slag. Store in a dry place.

Fluorspar, CaF_2, Fluorite, Calcium Fluoride, is natural calcium fluoride, pulverized. We will use this as a thinning agent in our flux. It is a neutral flux, neither acid or basic, and when used in conjunction with lime, or by itself, will help the slag separate from the metal (bullion) after the pour and the metal is cooling. Why pound on your gold until it looks like overworked brass? Get the flux right and the slag will snap free of the gold or silver bullion. Fluorspar is used as a flux in the steel industry, as a flux for metal smelting, in the manufacture of emery wheels, as a paint pigment, and is used in the optical industry. Ceramic supply houses carry Fluorspar since it is used in that industry as well. There are arguments, both pro and con, about using Fluorspar as a flux ingredient, however most smelters and assayers have been using it for many, many years with excellent results. Fluorspar is basically inert, and low in toxicity as a dry powder, but don't breathe the dust. Fluorspar creates fluorine gas at high temperature. Fluorine gas is lethal. Be careful. Fluorspar is not a primary ingredient in the slag. Store in a cool, dry place.

Bone Ash, as in calcined (roasted) bones. A coarse white powder. This is an inert, non-combustible material used to cover furnace floors in the assay business. A spill or boil over is caught (hopefully) by the bone ash to prevent damage to the furnace floor. Furnace floors are expensive. Fluxes will bore through firebrick and most other refractory materials. Every time you bring the furnace back up to temperature, the flux will reheat and continue dissolving the furnace floor. So, a layer of bone ash will catch the spill so It can be scraped out of the furnace. If you boil over a smelt, it would be really nice to have a layer of bone ash to trap the globules of gold or silver, which when cool, can be crushed and panned to recover the precious metals you would have lost. A quarter inch or so of bone ash is considered adequate. Never put bone ash in your flux. It is refractory, and will not liquefy in the smelt. You will have a mess on your hands. Store in a dry place in the original container.

Store ALL Chemicals Properly.

Potassium (or Sodium) Nitrate, Nitre, Saltpeter, KNO_3, is a transparent, colorless or white crystalline powder or crystals. It is sensitive to shock or heat, and should not come into contact with organic materials. Sodium nitrate is preferred, as it is pelletized and coated to keep the chemical from absorbing moisture. It can cause a fire if it comes into contact with organic materials. It is slightly hygroscopic, and will lump if not kept in a tightly closed, sealed container. If you were in the military, you should know what saltpeter does. (Not true, one teaspoon of saltpeter will make a hundred pounds of mashed potatoes so salty you can't eat them). Nitre is a strong oxidizer, and is used in the explosives industry, to manufacture solid rocket propellant, in the glass industry, in the tobacco industry, and for curing foods, as in nitrates. It has a low toxicity, but don't breathe the dust. We will use this in our silver smelting flux to check our slags for suspended silver. Niter is used in the fire assay of carbon, and high sulfide ores as a matter of course. It will oxidize impurities into the slag, and is used to control the lead button size in the fire assay. A very useful flux ingredient for assayers and those of us that smelt silver. Our oxidizer in the gold flux is manganese dioxide. Niter is not a primary ingredient in our flux. Read the MSDS and store accordingly.

Sodium Chloride, Salt, is a white crystalline powder. The salt mentioned here is *uniodized salt*, and is normally produced by refining halite, rock salt. The only reason it is mentioned here is that a few old timers use salt for a "crucible wash". A layer of salt is added to the melt and will be a clear, colorless layer that follows the pour out of the crucible. Some people argue against the use of salt. The theory being that as the chlorine portion gasses off, it carries values. As the gasses cool, the values accumulate in the ventilation system as high grade flue dust. Salt will also create a fair amount of smoke, some of which is chlorine gas. You can try salt as a crucible wash, clean crucibles are certainly nice, but you have to decide if there are losses from your smelt. Iodized salt is frowned upon because it simply introduces another unneeded chemical (iodine) to the smelting process.

Nitric Acid, HNO_3, Aqua Fortis, is a transparent, colorless or yellowish liquid that will attack most metals, except gold. It will dissolve silver rapidly, and will dissolve the silver in high silver, low gold alloys. Don't use it around silver or silver alloys unless you know what you are doing. It is manufactured primarily by oxidizing ammonia with air or oxygen, and can even be produced in nuclear reactors. Two tons can be produced from one gram of enriched uranium, they say, but this isn't the primary method of manufacture. **This is really nasty, toxic stuff!** It is toxic by inhalation, it is corrosive to the skin and mucous membranes, and is a very strong oxidizing agent. Wear a rubber apron and rubber gloves, face shield and use only with adequate forced ventilation. Never use except under a ventilated hood. If you add this acid to a mineral sample, the *red gas that evolves will kill you quick*, or seriously injure your lungs. That's why they call it the "red death". You may get better, if you survive, but you will never be well again. Try to buy this, and any other acid in plastic containers. If you drop a glass bottle and break it, the acid will splatter everywhere, the red fumes will evolve, and you will have horrible, disfiguring burns on your tender body. Every assay lab has one scarred veteran who didn't pay attention, or got careless, and paid the price. We will try to avoid using this, since it would be used to separate slag from the bullion, and if we do use it, it will be very dilute. Technical Grade is OK for our purposes. Read the MSDS.

Acids can burn you badly.

Sulfuric Acid, H_2SO_4, is sometimes used in dilute form to help remove sticky slags from gold and silver Dore'.Hot sulfuric will dissolve silver, but not PGM's, so assayers tend to have it on hand for Platinum Group Metal fire assays. Be very careful with this acid. It is incredibly exothermic, and will react violently if not diluted in the correct manner. **Never add water to acid!** Always add the acid to the water, remember A&W, as in root beer....Acid to water. If you pour water into a container of acid, it will react violently, and blow out of the container. The reaction is exothermic (creates heat).

Do you want your children around any of this stuff? No. Do you want your children or pets around a high temperature smelting operation? No. Use your head. Think about liability. Don't try putting this in your flux. The vapors from mineral acid are very corrosive, so store accordingly. Never let a bottle set around open, and remember that the dilute acid is also corrosive. Anything metal will begin to oxidize (rust) in the presence of the vapors. Read the MSDS, and store accordingly.

Check out Appendix A. You will find a complete sample MSDS, including a glossary, on lime. The MSDS is there to show you the quality and quantity of information that can be at your fingertips. Study the MSDS carefully, it can provide information that will prevent serious injury, or death. Be sure to ask for an MSDS, or a copy of an MSDS when you order your flux ingredients, or any other chemical.

Gary Vincent with good gold Dore' from his workings.

Chapter Four

Equipment And Fixtures

General:

Let's face it, you can't do the job without the tools you need. In some areas you can improvise, or make the tools necessary to do the job. First and foremost, you will need a heat source. This will be based on the volume you intend to smelt. If we are talking 75 lb. bars here, you will need a medium to large tilting furnace. If you are only doing a few ounces at a pop, a small, high temperature assay type furnace will do nicely. If you use a small furnace, it will have to be set on a table made of, or covered with metal or a refractory insulator, such as firebrick. Try to avoid wood when you choose your fixtures. The less that is flammable, the better.

Lay out the room arrangement so that the furnace and appropriate ventilation is separate from the area where you will weigh material, do paperwork, mix flux, or things that do not require close proximity to your heat source. Arrange your safety gear, fire extinguishers, first aid kit and such where they are easily accessible. Make sure you have at least two escape routes in case of fire. Unlock the doors before you start. You might be in a hurry if you have to leave.

Never have running water near your smelting operation! If you have to, build a wall and use the thickest drywall you can afford to separate the wet area from the smelting area. Drywall with a sheet metal covering is a great insulator. Top your tables with it, or your workbenches, but seal the edges with a molding to prevent damage to the drywall, or bend the sheet metal over the edges to protect the drywall.

Use secure premises. You don't want your kids, the neighbor kids, nosey neighbors, pets, or anyone else wandering around while you work. The risks are just too great. Trust me, this is the voice of experience you are hearing. Remember the magic word...Gold! Got any burglars in your neighborhood? If you don't have them now, you will when the word gets out. Crooks think every assayer alive automatically has a personal stash of gold hidden on the premises. (Huge yellow bars, no?) None of this is worth dying for, so don't. Think it through, be very, very careful who you talk to, and make sure your spouse and children either don't know what you're doing, or can keep their mouths shut. You will soon find out who your friends are.

Restrict access to your operation.

If you can't weld, find someone who can, at a reasonable price. You can build, or have built, very nice tables, or benches from inch and a half square tubing. If you have an oxygen acetylene setup, you can bend your own tongs, scrapers, molds, and other handy implements. The oxygen acetylene rig will become important when you coat your mold, as you will see later on.

Attitude Towards Safety:

Have a good attitude about safety. Think through the "what ifs", and be prepared, like a Boy Scout. You will have problems sooner or later, so think it through. What will you do if you are in the middle of pouring a smelt, and slag drops on the floor and a fire starts? Did you think about where to set or hang the fire extinguisher? What if you drop a hot crucible, and splatter superheated metal on your leg? Where's the first aid kit? Are you wearing lace up boots? Tennis shoes? If you are, you are asking for trouble. Think it through, no amount of gold can compensate you for serious injury, or death. What about insurance? If you burn your building to the ground, will your insurance pay for the damages?

Keep water away from high temperatures.

The Basic Structure and Fixtures:

You should have your basic structure picked out, the ventilation installed, tables and counters made of a fireproof material, the water facilities (bathroom, etc.) isolated from the smelting area. If you are using an electric furnace, the power should be ran to the appropriate location. You should have two exits, both unlocked when you are working. Have your fire extinguishers in place. Don't use a sprinkler system. If it triggers, and sprays water on a 2100°F furnace, you will have a serious explosion on your hands. A burglar alarm is useful, and if you install a safe, use a floor safe that is concealed from view. Don't use a fire safe. They are a burglar's delight. Any vehicle with a winch will pull it right through a wall, or window. Remember that security will be important, whether you have gold on the premise or not. The bad guys will know you have gold on the premise whether you do or not, and will come after it.

Personal Protective Equipment:

As previously mentioned, you should have all your safety gear available. You should have the protective clothing described earlier, a face shield, preferably the gold coated assayers face shield, a pair of safety glasses and goggles, a respirator, a pair of heavy leather work gloves, a few dust masks from the hardware store, and some latex rubber gloves from the drugstore. If you have it, you'll use it. Respirators and face shields should not be stored in the same area as your smelting equipment. The dust that settles in the room will probably be metal oxides. Keep your respirator clean. Disassemble and clean it at least once a week.

Heat Source:

As mentioned in the previous paragraph, you will need a heat source, capable of sustaining 2150-2300°F continuously, 2500°F intermittently. That means that a ceramic kiln will not work. Don't waste your money, buy the right equipment the first time.

How much precious metal do you really expect to produce? We would all like to think we will pour a couple hundred pounds with each smelt. The reality is usually a lot lower, say ten to thirty ounces at a pop. So, instead of trying to impress someone, be realistic. Using a decent electric assay furnace with a controller, you can easily smelt gold. A high quality assay crucible will work, but the high temperature will shorten the life of the crucible. The crucibles are fairly inexpensive. Look for the DFC Colorado or A. P. Green 40 gram assay crucibles. Either will smelt 30 ounces of gold.

Also keep in mind, a small electric furnace as described will allow you to "pilot test" your flux with the metal you produce. Your first pour should pay for the furnace. If you need to upgrade as your operation increases in size, it can easily be done.

Electric Furnaces:

There are some electric furnaces that simply won't work.. A Scut ceramic kiln will heat to cone eight, hot enough to assay, but not hot enough to smelt. There are other electric furnaces that will easily reach assay and smelting temperatures, notably the Vcella kilns. These kilns are getting more and more popular as people realize that temperature control is necessary. A Vcella kiln will hold 2150°F all day long, and can reach 2500°F intermittently. Vcella kilns are reasonably priced, easily rebuilt, and spare parts are very reasonable. Action Mining sells a Model 13 Vcella with their fire assay kit.

Vcella's TL-60 Tilting Furnace.

Vcella also makes tilting furnaces in several different configurations. The other important thing about Vcella Kilns is that *you can get a very accurate furnace controller for any model*. You can control the temperature of the smelt, there is no guesswork at all. Note that your author has poured many, many hundreds of assays and smelts utilizing Vcella's Model 16 furnace, and developed the information you are reading with a Vcella Model 60 tilting furnace.

The one thing you have to know is the line voltage before you order. You should know the line voltage for any electric furnace. If you order a 220 volt furnace, and your line voltage is 208 volts, it will take four or five hours to heat to 2000°F. If you don't get the right match in voltage, the elements have to be replaced. Furnace controllers are an option, and must be ordered separately.

There are now other options out there, as illustrated by the photos of the RapidFire Kiln. The web page, www.goldrushtradingpost.com, says "RapidFire Pro-L comes complete with a digital control panel, 1 alumina-fiber shelf, embedded heating element, 6" x 5" x 6" firing chamber, 1500 watts of power, 1 set of kiln legs and wide opening door. At just 11 pounds, it's light and portable. Also available with programmable ramping controllers, and in a variety of colors." It's an inexpensive way for firing a small quantity, say up to 10 ounces in an assay crucible. Check it out. Note that the author has never used this apparatus, and can't endorse the product. Your choice. And thanks to H. Geiger for the photos.

Furnace Controller:

A RapidFire Kiln. A Rapidfire kiln at temp.

Very handy. Normally, a furnace does not come equipped with a controller. It does just as the name implies, it controls the temperature of the furnace. Set it where you want, and go do something else while the furnace heats up. When the furnace reaches temperature, the controller will keep the furnace at that temperature, plus or minus a few degrees. Notice the controller in the Vcella photo on the previous page. It is at the top right, sitting on the framework. Controllers start at $500 and go up, way up. Best buys we've seen were at Vcella Kilns and now they are all digital.

Induction Furnaces:

A small induction furnace.

Induction furnaces are very fast and efficient. They are expensive to purchase and set up. Induction furnaces usually require a lot of voltage to operate the larger furnaces. The interesting thing is that the electrical eddy currents create a swirling, or stirring action in the crucible. A large generator is always a good option to consider. Induction furnaces can easily reach and exceed normal smelting temperatures.

The drawbacks are initial cash cost, installation cost, and operating cost. Regular maintenance is a must. The crucibles normally used are silicone carbide, and are sealed around the top when placed in service. The pumps, coolant reservoirs, plumbing and electrical circuits also require periodic maintenance. A close examination of the crucible after each smelt is a must. Crucible failure that allows the flux and or molten metal to contact the liquid bearing coil can be catastrophic.

The quick heat appears to evolve the oxygen in the flux much quicker than a typical electric furnace, so the operator should not attempt to smelt any material or concentrate that is not extremely clean. A placer concentrate with substantial black sands, or the direct smelt of "high grade" ore will

A pour with a larger induction furnace.

yield some spectacular iron alloy compounds, that are just about impossible to break down to a Dore' acceptable to a refiner.

In any case, it is not cost effective to refine the iron out of the pour chemically or otherwise if you are a small or medium operation. The disposal of the spent chemicals can be very expensive.

Muffle Furnaces:

A muffle furnace typically is a silmanite shell surrounded and heated by electrical elements, or a gas burner that provides the heat. The purpose of the shell is to insulate the material being heated from the high volume of air the gas blower produces. In the electric furnaces, the muffle insulated the material being heated from contamination.

Historically, gas-fired muffles were used in the fire assay process. The limitations were low capacity, and had no temperature control. The newer units have solved those problems, but most mines utilize electric furnaces built by companies such as DFC (Denver Fire Clay) based in Canon City, CO.

As with most other furnaces, parts are very expensive for muffle furnaces.

Microwave Oven

Here is another cheaper way to smelt. Go to www.microwavegoldkiln.com and watch the U-Tube video. You will notice a few things about the video, the first being no comments on ventilation or other safety issues. The other is a "proprietary" or special flux. The main thing is to notice that the temperature was low, and there is no mention of the duration of the heating process. The crucible appears to be a standard 40 gram A. P. Green assay crucible. The crucible is insulated with a very light firebrick material, such as that used in Vcella Kilns. Do your research, and if you think this will meet your demands, well, go for it. Never, ever use the microwave for cooking food after this.

The crucible charge appears to be 30 or 40 grams, flux ratio three or four to one. The manufacturer should provide all the information with the purchase of the kit. Also read the caveats on the web page. Honestly, if you have too much black sand in with the gold, you will still create a separation problem. Think *any* method that you use all the way through. Note that the author has never used this apparatus, and can't endorse the product. Again, your choice.

Gas Furnaces:

This is probably the most common furnace used in small scale operations. The problem is that these small, round furnaces have the gas ported in the side, which is ignited. The crucible is then placed inside. The expanding gases exit the top of the furnace in a circular manner, which will suck the goods right out of the crucible. The expanding gasses create a low pressure area above the crucible. It's pretty depressing to watch, even more depressing when the operator sees his fine gold disappear into the sunset.

Always have a cover for the crucible used in this type of furnace. At the very least, use a good borax glass caps on the smelt.

Button from dirty con, low heat. (Photo by H. Geiger)

Normally, there is no temperature control for this type of furnace. Temperature control is provided by the operator's magic eyeball. Here's a hint...Try to hold the furnace at a straw yellow interior until the fusion is quiet. No bubbling, and not so much smoke. Or better yet, get an infrared temperature sensor described below.

Also remember these aren't "fire and forget" furnaces. You have to monitor them constantly from start to finish. Don't fire the furnace and walk away. That's a nice way to find your metal dripping out of the bottom of the furnace when your crucible burns through because you're running the furnace too hot.

Don't even think about using this type of furnace indoors without some serious ventilation. Also remember they sound like a jet taking off. Always put a half inch or so of bone ash under the crucible, in case of a burn through.

Infrared Temperature Sensors

For those of you that wind up with a gas furnace, or an induction furnace, you can use a high temperature infrared sensor. These devices are handheld, and the low temperature ones are dirt cheap. The higher the temperature, the higher the price. They are very easy to use, and solve a lot of problems. The knowledge is certainly useful. Many people have thought they were at smelting temperature, only to find out they were barely at assay temperature after testing with this device.

If you aren't completely familiar with the various colors of furnaces, get one. What one person perceives as straw yellow can be different to the next guy. Temperature control is important!

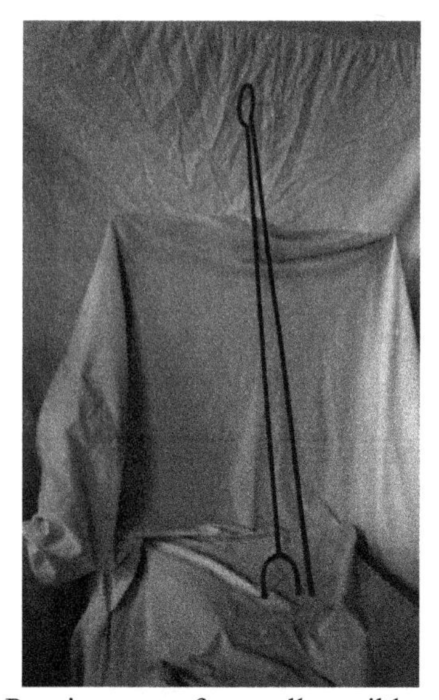

Pouring tongs for small crucibles.

Tongs, Scrapers, and other handy items:

Make these yourself, or head for the nearest welding shop and have them bent up and welded for you. Never order them from an assay supply house, you will pay over $100 for $9.00 worth of ⅜ rod. You also won't know what size you will need until you have the crucibles you will be using.

The 'U' of the tong should come ⅔ of the way up the outside of the crucible, and have a safety bar that comes across the top, or mouth of the crucible to hold the crucible in the tongs so you can completely invert the crucible. A scraper for the furnace floor is made by simply welding a piece of angle iron perpendicular to the end of a piece of ½ black iron pipe four feet long. Nothing real difficult about this. See the photos on the previous page if you don't understand this description.

Balance, or Scale:

This is a must. You will need a balance to weigh your flux, to weigh whatever you are smelting, and to weigh the finished product. Get a set of weights to check the calibration of the balance while you are at it. A new dollar bill will weigh one gram. But that doesn't do you any good in the kilo weight range, does it? Work in grams. It will make your life a lot simpler. The conversions are easier. Forget pennyweights and grains. These days, it's grams. Learn the conversions to the troy weights, and you're on your way. Your balance should have a capacity of 2500 plus grams, and a readability (accuracy) of .01 (as in one hundredth) gram. A tenth of a gram readability is OK for mixing fluxes and such, but not accurate enough for weighing the noble metals.

A Sartorious Balance

And here's why:

One troy ounce weighs 31.1035 grams. Each gram, at $1200 per troy ounce is worth $38.58 (divide $1200 by 31.1035). Each tenth of a gram is worth $3.86, and there are a lot of tenths in a troy ounce. Each hundredth (.01) of a gram is worth $0.38 or 38¢. We all can pretty much live with the 38¢ error, but a $3.86 error is considered an unacceptable error when working with noble metals.

So pick out the right balance from your lab supply house, or chemical supply company. A good balance will cost about $500.00 these days, but is well worth the price. Assayers use bead balances that are 10^{-3} gram, and cost about $7000 and up. These are micro balances, and we won't need one to do our smelt. Reloading scales in grains do not have the capacity you will need, and the platform scales are not accurate enough to weigh your bullion. Take a look at the balance in the photo to get an idea what we are discussing here. It is a Sartorius balance, and was bought used for $300. Works great, has a capacity of three kilograms, (3000 grams) and a readability of .01.What a deal! And they are out there.

Also be sure to get a calibration weight to test the balance with, and do not handle the weight with your bare hands. The natural skin oils in your fingers will corrode the weight, or cause the weight to corrode. Wear cloth cloves when handling the weight.

Pouring Molds:

A variety of pouring molds are out there. Conical cavity assay molds, ingot molds, huge conical slag pots on wheels used by the mines, refractory meehanite molds, adjustable sliding bar molds, graphite molds, and the homemade molds made from angle iron. Any and all of them work from one degree to another. The most important consideration, first, is the size of your pour. If you intend to pour small one ounce buttons, which are easily marketable, a conical depression assay mold will do nicely. A larger cavity version is also available. (See illustration) The conical depression helps the molten metal settle to the bottom better, and some people claim It will help the slag separate better since the slag and the metal will cool at different rates. Makes sense. The regular ingot molds (see illustration) come in sizes from one ounce to one thousand ounce. The ingot mold (rectangular) in the illustration will hold five hundred ounces of gold, and was purchased from DFC Ceramics. **Never pour to a bare metal mold! Always coat your mold with a release agent!** Superheated metal will weld to the bare metal, and you will be hating life as you try to get it loose. There are a lot of graphite-based mold releases on the market. Any company dealing in refractory products will have mold releases. A handy thing to use is acetylene smoke from your oxygen-acetylene cutting or welding torch. Leave the oxygen off, crack the acetylene valve, light the torch, and let the acetylene smoke (carbon) coat the mold as you move the tip around the inside and outside of the mold. This is the "poor boy" method, messy, but It works well. Do it outside, or under a hood, or you'll have acetylene smoke settling everywhere. The only disadvantage is that you will have to wash your metal, and normally, most of the (carbon) smoke comes off when you are repeatedly quenching the metal. A little hot soapy water and a scrub brush will clean it right up. If you make a mold out of angle iron, cut the ends of the angle iron at a 45 degree angle toward the center, so you won't have straight sides. Use quarter inch or thicker angle iron. Weld the seams inside, and do a clean job so no protrusions are inside the mold for the metal to hang on. Four triangles welded together with the point down will make a nice slag pot, or pouring mold. (See illustration.) You can have a metal shop roll up a conical mold. Stay with plain old steel,

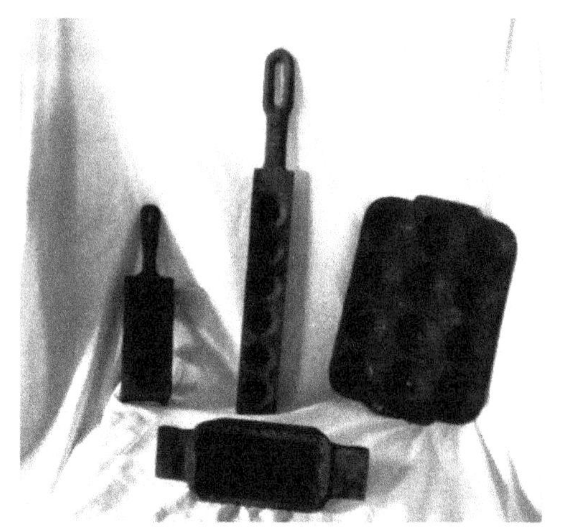

Pouring molds-3, 6, & 12 depression conical at the rear, 500 Oz ingot mold in front.

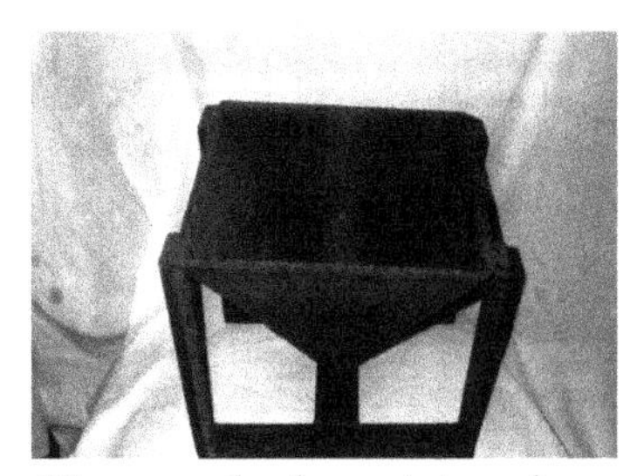

Home made slag pot, top view.

Home made slag pot, bottom view.

whatever you use will have to be coated with a mold release. If you need to catch excess slag under the mold, an eighth inch, or 11 gauge sheet metal tray is easy to make.

Odds and Ends:

A large slag pot with wheels.

Metal and plastic scoops are handy when weighing flux, or making flux. A set of measuring scoops and spoons are also useful. A clean metal bucket to fill with water for quenching the metal is a good idea, just keep the water away from your heat source, or electricity. Forceps, as in giant tweezers, a foot or so long, are real handy. A counter brush and dust pan are a must for sweeping up slags. A good push broom and straight broom are a necessity, especially if you are using a tilting furnace. A ball peen hammer is handy for fracturing the slag if it doesn't shatter as it cools. A hot plate or old stove is handy for drying cons, precipitates and other things you wish to smelt. **Never smelt wet or moist material!** Steam will evolve and cause you to loose the material, or if trapped under a molten cap, will explode. A 35 gallon oil drum with a cover, or a metal bucket with a cover are necessary to store your slags. If you need to separate batches of slag, more than one container will be required. It is customary to return the slags, crucible, and metal if you smelt for someone else, so think about keeping everything separate. You might want to keep one particular batch separate for assay purposes. Don't leave the slags exposed to air if you intend to have them assayed. They are hygroscopic.

The Notebook:

One of your most important tools. Make sure it is hard bound.

This is the single most important item in your shop. **Write down EVERYTHING!** There is nothing more frustrating than trying to remember how you compounded a batch of flux, what the precipitates you smelted weighed, and so on. Record all your recipes. Suppose you smelt a 500 gram sample of gold dust, and the bar (bullion) weighs 250 grams. This will tell you the material is about 50% gold. Useful information. Let me say it again....**Write down EVERYTHING!** If the IRS comes after you, they will want to see a "bound" book. We'll get to the IRS later in the book. So don't use loose leaf binders, go to the office supply and get a bound record book, or go to a lab supply house and get a regular lab notebook. Always use the pages in order, never leave blank pages as you write. Always write down the date, time, your name, and step-by-step, what you did. Guard the book, don't let it get away. Don't use any fancy codes or anything. Just secure the book in a safe place until your next session. Never loan the book out, or let your buddies have access to it. You'll find pages torn out, or the book will disappear. Think of all the great secrets you will write about! Think about being able to go back to your book and know exactly how you mixed the last batch of flux. For troubleshooting purposes, your record book will be the handiest thing you have.

The book can be used to settle accounts, for arbitration, and even as evidence in a court of law. Take care of your notebook.

A pour to a home made slag pot. The point is mostly silver.

Chapter Five

Fluxes

General:

A "Flux" is defined in the dictionary as a substance used to refine metals by combining with impurities to form a molten mixture that can be readily removed. Is this going to be fun, or what?

What we will be doing here, in one case, is modifying a gold smelting flux that has been around for a long, long time. For many years, the United States Mint could produce gold at a fineness of .999, while mints in other countries could produce bullion that was only .900 fine. The secret ingredient was the manganese dioxide, which has an affinity for silver, and causes the silver to go to the slag, or "slag off". Silver, and all other base metals, such as copper, lead, zinc, iron, etc. are considered "impurities" in a gold smelt. The manganese dioxide will also slag off the other base metals previously mentioned, thus causing the remaining gold to be of a higher purity, or fineness. Obviously, it would not be a good idea to smelt any material with appreciable amounts of silver with a flux containing manganese dioxide, since the silver would be lost to the slag, making recovery of the silver difficult.

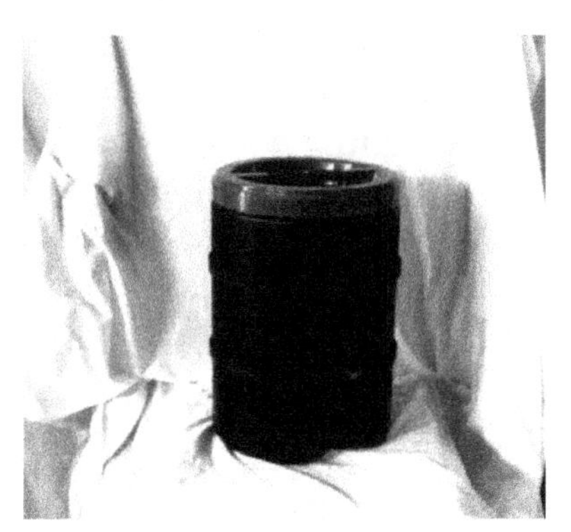

The Oddjob mixer. Great for mixing flux.

Manganese dioxide will give your slags a beautiful, deep purple color. If you use less than specified in the flux recipe, the slag will be a lighter shade of purple. The silver flux, which can also be used for gold, typically will give you a light "apple green" slag. The silver flux is used on gold-bearing materials when the silver is to be kept in the bullion. If you have a concentrate containing a preponderance of silver, and a small amount of gold, this is when you would use the silver flux on a gold-bearing concentrate. Or the silver could be separated from the gold with various wet chemical techniques, then smelted to bullion with very little gold remaining as an impurity.

Normally, the gold contained in silver bullion does not bring the dollar amount that is desired, since some expense is required to separate the two elements. So you will have to decide what you want to do with the material before you smelt. Silver, at $15 or $20 dollars an ounce, is not considered economical to separate unless you have a well equipped lab set up, and ready to go. In this case, the silver is slagged off with the gold flux, and considered part of the expense of refining. If you part, or chemically separate the silver with nitric acid, the acid is expensive, and a fair amount of acid is required to accomplish this. Also, a lot of the platinum group elements, if present, are soluble in the acid, and will be lost. Do your homework. There are better methods out there.

Inside of an Oddjob mixer. Note mixing paddles at top and bottom.

The fluxes we will mix, or variations of these recipes are used by mining companies all over the world. They may use more or less of one or another ingredient, but are fairly consistent in their recipes world-wide.

There are a few things you need to remember about the flux recipes and smelting:

Never add a reducing agent to your flux! Reducing agents are sources of carbon in the smelt, such as flour, sugar, cyanide, sulfur, or carbon. Carbon sources, such as flour, will cause the base metals and other impurities to reduce from the slag and contaminate your bullion. Iron can be particularly nasty in a smelt, and the addition of a reducing agent, at smelting temperatures, will create some iron compounds that will require chemical pretreatment of the bullion, or successive smelts, to remove the iron compounds from the bullion. Remember that we are trying to oxidize the impurities to the slag, not reduce them to the bullion. Cyanide will gas off and kill you quick, even if you have serious ventilation. ***Never, ever use cyanide as a reducing agent!***

Never add lead in any form to your smelt! Some amateurs will add litharge, red lead oxide, granulated metallic lead, lead foil or lead wool to their smelt, and pour a lead bar. This will require cupellation to separate the precious metals from the lead, and just adds another step after the smelt. Lead also creates toxic waste, so let's avoid lead at all costs.

Why work harder? Read more about industrial cupellation in Chapter Eight. If your material won't smelt as it is, upgrade it (re-concentrate) or pre- treat with chemicals to upgrade the quality of your material. Don't contaminate your material by adding lead. Assayers use lead as a collector in a fire assay. Lead has an affinity for silver, which has an affinity for gold. We are not assaying here, we are refining a high grade product to a marketable form. We should be far, far beyond assaying at this point.

The methods in this book were used to smelt the gold & silver bars in this photo.

Never add base metals, such as copper, to your smelt! Again, some people are convinced they need to take about five extra steps to accomplish what we will in one step. These people will insist that the copper, or whatever, will be necessary to enhance, or boost the recovery of precious metals. Bull. All that the addition of copper or other base metals does is create extra work, and extra reagent costs that will easily offset the hypothetical gains in precious metal recovery. If you subscribe to this theory, try small batches of 30 grams or so, and calculate what your time is worth, and what the chemicals cost, and you will soon come to realize that you have increased your overhead and effort for the same return. Not smart.

Always have at least fifty (50%) percent metal available in your material! If you mix your flux and material to be smelted according to the directions in the recipes, smelt, pour, and have no metal to show for your efforts, you didn't have enough silver or gold in the material to act as a collector in the smelt. Save the slags for later, re-process your material, and try again. If you think you can take a rock that has an ounce or two per ton assay, smelt a ton, and recover the ounce or two, you're wrong. You will spend a small fortune doing this, and it won't work. Direct smelting of an ore is done only when it is of a very, very high grade, and a metal in some form is available as a collector. Make sure the ratio is correct before you start. An assay is all you need to tell you whether or not your material is of sufficient grade to smelt.

Making The Fluxes:

Well, here we are. You have your chemicals (reagents) on hand, you've read the MSDS's, you have the necessary safety equipment and a well- ventilated place to mix your reagents. You want to mix up enough flux for both gold and silver, say a gallon jar full of each, to start. We will mix our ingredients by weight, mix them well, add one part of our material to two parts of the gold flux, and we're ready to smelt.

Note that we are mixing by weight, not volume. In other words, we use a unit such as pounds when we mix our flux, not how full the ingredients fill a beaker, or other container. Remember that we will be using technical grade chemicals, not the more expensive reagent grade.

A useful container for fluxes are the large plastic jars that restaurants use for pickles, mayonnaise and other condiments. Normally, these jars are thrown away. Ask for them, and they will give you all you want, free. The same for plastic buckets. Make sure you get airtight lids for the buckets and jars. Take the jars or buckets home, wash them out thoroughly, dry them very well, and you're ready to mix fluxes. **Make sure your containers are dry!**

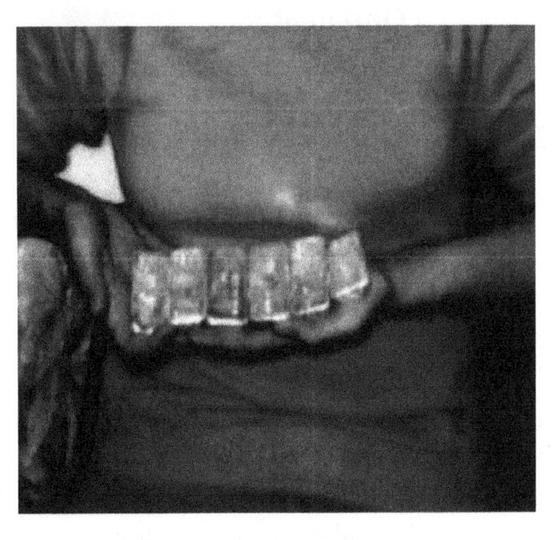

Silver produced using the methods and flux in this book.

The Original Mint Flux (Gold Only!)

Here's the recipe for the original mint flux that started it all:

Three (3) parts Borax Glass
One (1) part Soda Ash (Sodium Carbonate)
One (1) part Silica
One (1) part Manganese Dioxide

Read on before you mix any of this flux. The reason this recipe is provided is that you may encounter a material to smelt that is very fluid in a pour, and has fair amounts of calcium as an oxide present. If so, this flux, as is, can be useful to you. The problem with this formula is that the large amount of borax glass in the formula will cause the slag to seriously stick to the bullion. The slag is very viscous, and tends to retain metal, or "shot" the slag. The thick, viscous slag will also allow the metal to run past the slag, during the pour, and spatter, resulting In a loss of metal. Serious work with a hammer is generally required to get the slag off the bullion. After prodigious pounding to remove the slag, your gold bullion will look like reworked brass.

Let's solve the problem:

Gold Smelting Flux (Gold Only!)

(Use Two Parts Flux to One Part Material Being Smelted)

Furnace at 2150°F for at least 30 minutes **after** reaching temperature.
Cap smelt with a **thin** layer of Borax Glass.

Two Parts Borax Glass (Anhydrous Borax, $Na_2B_4O_7$, Glass)
One Part Sodium Carbonate, Soda Ash (Anhydrous, Na_2CO_3)
One Part Silica (SiO_2 Silicone Dioxide, Sand)
One Part Manganese Dioxide (MnO_2)

Optional:
One Half Part Fluorspar-Lime Mix (See below)

Lime-Fluorspar Mix (Use sparingly!)

One Part Fluorspar (CaF_2, Calcium Fluoride)
One Part Calcium Oxide (CaO, Lime, Type S, Unslaked)

If you compare this to the original Mint flux formula, you will notice we have cut back on the borax glass, which caused the slag to stick so badly. We have added a couple of chemicals to increase the fluidity of the smelt, namely the lime and Fluorspar. Both are thinning agents. The added benefit of the lime is that it is mildly basic, and a weak oxidizer, that will help prevent the slag from sticking to the metal.

Let's make the lime-fluorspar mix first. Weigh up a pound of Fluorspar, put it in an appropriately sized glass jar that has been labeled "Flux Mixing Jar". Weigh up a pound of lime, and add it to the jar. Cap the jar, and using a turning motion, **mix the two ingredients until they are the same color.** Usually, a couple of minutes of mixing will do the trick. Log the flux in your notebook, include the date, time, location, and any other notes you feel are necessary. Put the lime-Fluorspar mix in an appropriate storage container, and label the container. Include the date when you label the mixture.

Bear in mind that the fluxes can be mixed in any mixer. The only catch is that you want the mixing device to be air tight. If you are making really large batches, use a cement mixer after slowing it down ten RPM (change the pulleys). Cap the mouth of the mixer with a large, heavy duty trash bag and a bungee cord like the truckers use. Let the dust settle for ten minutes or so before you pull the cover off. For small batches, a rock tumbler can be used, except that a gallon jar usually has just as much capacity, and the glass walls allow you to see the color of the mixture. Then you know when you're done mixing.

Now, we will mix up the main flux. Put one pound of Manganese Dioxide in your jar. Manganese dioxide is very dense, it won't take much to make a pound. Add two pounds of borax glass, one pound of sodium carbonate, one pound of silica, and one pound of the lime-fluorspar mix to the jar, leaving a little air space so everything will mix. Turn the jar until everything is the same color. Voila'! Gold smelting flux! Place the flux in an appropriate air tight storage container, label it clearly with the flux type, date, and any other information you feel is appropriate. Record the weights and other information in your notebook, so you will know all the pertinent information the next time you wish to make a batch. How many pounds did the formula yield? How much volume do you have? This is what the notebook is for.

Never mix your gold or silver bearing material with the flux in the flux mixing jar. This will contaminate every batch of flux you make. **Use a different container to mix the flux and gold or silver bearing material!**

Be sure to label the container as a mixing container for the concentrate, or whatever you are smelting, **not** as a flux mixing container. Record any batches mixed in your notebook, and label the mixture.

Gold Smelting Flux #2 (For Smelting Placer Gold)

(Use One Part Flux to Three to Five Parts Gold)

Furnace at 2150°F for 30 minutes **after** reaching temperature.

Cap smelt with a **thin** layer of Borax Glass.

Three Parts Sodium Carbonate (Soda Ash)
One Half Part Borax Glass
One Part Silica (Sand)
One Part Manganese Dioxide
Optional:
One Half Part Lime-Fluorspar Mixture

Notice the similarity to the previous recipe. This recipe is for placer gold, or other very high grade gold-bearing material that is at least 70% gold. It will slag silver, and is a very good formula for re-melting gold from the first formula that has iron, copper, or other contaminates that have carried through the smelting process. You can use this flux as a substitute for the first formula in a pinch, just reverse the ratio and use three parts flux to one part material.

Make this flux using the same techniques described for the first formula. Be sure to label the flux accordingly, store in an airtight container, and make the appropriate notes in your notebook.

Now we will make a batch of silver flux. This flux is also useful for gold, especially when it is alloyed with large amounts of silver.

Silver Flux:

(Use Two Parts Flux to One Part Material Being Smelted)

Cap the smelt with a **thin** layer of Borax Glass. Furnace at 2150°F for at least 30 minutes ***after*** furnace reaches temperature.

Two Parts Borax Glass
One Part Sodium Carbonate (Soda Ash)
One Part Silica (Sand)

Optional:
One Half Part Lime-Fluorspar Mix

Option #1: One Part Potassium (or Sodium) Nitrate, KNO_3. (**Do Not** add to flux when mixing.)

Make and mix the flux as previously described. Label and store the same as the other fluxes. Make your notes in your notebook.

Use this flux on materials containing recoverable silver. This flux will not slag silver. You can use this flux on about any mixture of gold and silver you will run across. If you have problems with base metal contamination, you can add one half part of potassium or sodium nitrate as an additional oxidizer. Do not add the niter when you mix the flux. Add it if you need to when you mix the flux with the material you are smelting. The optional niter would be a very useful flux addition when smelting sulfide concentrate, since the sulfur is a reducing agent in the smelt. The niter, as a strong oxidizing agent, would prevent the sulfur from reducing base metals such as iron, in the smelt.

The fluxes you have made can be used in any size smelt. Take a pinch of the flux, a half pinch of your material, mix it with a match stick, put it on a piece of angle iron, and heat it gently with an oxygen-acetylene torch. Use the flame to stir the molten mixture, let it cool, and check the results with a 10X magnifier. You should see metal balls in the slag. This is a quick test to see how the flux will react with your material.

Pouring a small smelt using assay equipment.

Most people who smelt will run a small 15 to 30 gram test smelt to see how the flux will react with the material to be smelted. This is done by using assay size pouring molds and assay sized crucibles.

If you are smelting cement silver, or silver chloride, never use an assay crucible. The silver will weep through the sides of the crucible and onto the furnace floor. Use a fused silica crucible instead. Mix the flux with the dry silver chloride, and hold at 700 to 800°F for two hours, then bring the furnace to temperature and hold as indicated with the flux formula. This will allow the chloride portion of the smelt to gas off as chlorine, which is extremely hard on furnace elements, or anything else it comes into contact with, including your lungs. You would be far better off to cement the silver chloride before smelting. The end result will be much easier to obtain, and you won't damage the equipment. The best mold to use for casting silver is a graphite mold. They are available from the supply houses listed in Appendix C, or you can mill your own out of solid graphite, and hand sand to a smooth finish.

A concentrating table is an easy way to clean up concentrate.

Chapter Six

Slag

General:

Does this seem to be an odd title to a chapter? Actually, the slag from your smelt will tell you how you are doing. A skilled assayer or person who is smelting will know how well they adjusted the flux, and have a good idea what elements are present in the ore from checking the slag from the pour. There are several characteristics that we are interested in, first and foremost being the silicate degree of the slag.

Appearance:

Our slag, when cool and separated from the metal we have poured, should be glassy in appearance. In other words, the slag should look like broken glass. If the slag has a dull, "stony" (as in rock, or stone) appearance, and pours like water, we are lacking silica, borax glass, or both. Dull, stony slags can also be from excess base, or alkaline ingredients in the flux. Typically, assayers have a two to one mixture of silica and borax handy to adjust the flux when needed. The stony slag does not necessarily mean that the smelt was not successful, just that the flux needs to be adjusted, or balanced out. If the slag is holding globules of metal, or poured out of the crucible very thick or viscous, separate any large pieces of metal from the slag. Crush the slag, place the metal, slag, and extra borax and silica back in the crucible, and furnace it again. Usually, the amount of silica and borax mixture added to the smelt is 10% of the original amount of flux. If you used a pound of flux the first time around, add a tenth of a pound of the silica/borax mixture for the second try.

Wear goggles, safety glasses, or a face shield when working around slag.

If your slags are very glassy, and stick to the metal you have poured, they are indicating an excessive silicate degree (too much silica, or borax glass) and are considered an "acidic" slag. Adjust the other way by adding soda ash or lime. If the slag was thick and viscous, the addition of the lime-fluorspar mix will help. Try adjusting in one tenth, or 10% increments.

What we are after is a glassy slag that shows no particular affinity for the metal, and fractures away from the metal easily. Some of the slag may stick to the metal as a function of temperature, but will separate quite easily when cool. When molten, the slag should pour like warm 30 weight motor oil. Pick up a piece of cold slag and check it with a magnifier, or under a microscope. There should be no visible particles of metal in the slag. The slag should pour into the mold ahead of the molten metal, and coat the mold before the metal begins to pour from the crucible. The metal is quite a bit heavier than the slag, and will pour last. Residual slag will follow the molten metal into the mold, and you should try to get all the slag out of the crucible that you possibly can.

Never throw the slags away. They may contain values, and in some cases, are crushed and used again. If you are going to dispose of the slags, assay them, or have them assayed so that you know what values, if any, are carried in the slag. There have been small fortunes made on smelting slags, so make sure you know what you are doing when you dispose of them.

If there are residual values trapped in the slag, the values can be recovered by crushing and pulverizing the slag, and using a very efficient gravimetric device, such as a refining table, to separate the values. Wet chemical methods can be difficult due to the base metals that have been oxidized into the slag.

If you smelt silver, and your slags are tan or brown, crush the slag, take about 30 or 40 grams of slag, add a teaspoon full of potassium nitrate, and furnace at 2100°F for two hours after reaching temperature. Pour the contents of the crucible into a mold, and when cool, check for metal. If it was in the slag, it will be in the mold. The tan or brown is usually caused by silver chloride in the smelt.

Color:

The gold smelting flux will give you a purple to purple black slag, as indicated previously. If you use the silver smelting flux, and see purple, you will know manganese is present. Cobalt will give you a beautiful blue slag, like the old time cobalt glass seen in antique stores. Too much silica will give you a thick, viscous coke bottle green slag. Lead typically will produce a lemon yellow slag. Tellurium or selenium will create a cherry red slag, and copper oxides will produce the classic turquoise colored slags. Iron produces a brown to coal black slag, antimony a yellowish green slag.

Smelting slags of different colors. Black (bottom left) indicate iron, Brick red (center and top right) indicate copper, bottom right, manganese (dark purple) top left is low copper (oxide) in silver smelt.

The thing to remember about the color of the slag is that it will tell you which element is present in the greatest amount, or preponderant in the smelt. You will not see a little blue here, a little green there, and some purple. If you are using the gold smelting flux, purple is all you will get, since the manganese will be the preponderant element in the smelt.

Hazards:

Slags will contain all the impurities in the material you smelted, and can certainly be toxic if ingested. Frankly, it is difficult to eat glass, but if the slags are hygroscopic, and most are, they will decompose in your dog's water dish. Sweep the slags up after they have quit spalling and cooled. Put them in a metal container with a lid, preferably air tight. People find slags to be "pretty", and will pick a piece up when they see them for the first time. Remember that slags will spall violently hours after being cooled, and are as sharp as glass, or sharper. Slags will cut like a razor, as you will find out if you work around them any length of time. Never handle them without good, heavy leather gloves, and always wear eye protection when you are smelting.

You are waiting for your first smelt to cool. Suddenly hear a loud pinging sound, and small things bouncing off the ceiling and walls. The slags are spalling. You realize what would have happened if you were bending over the mold with your face a foot or so away. When the slag has cooled to hardness in the mold, and still hasn't fractured, tap the slag with a hammer to fracture it. The smaller pieces may still spall, but the reaction won't be as violent.

It is socially acceptable to crush and reuse slag in place of flux, up to 50% by volume, **on your own material.** Never reuse slag someone else's smelt. This is considered uncouth, and not bright idea, considering the values that can be retained in slag. Yes, you can save a few bucks on the cost of flux, but you can also come up with strange results. If you are smelting for someone else, strange results are not what you are after. When you are messing around with other people's gold, stunts like that can get you seriously injured, or who knows, they might smelt you. Pay attention.

Always record the results of your smelt in your notebook. Record the color of the slag, whether or not it had an affinity for the metal, what it poured like, was the slag clear and glassy, or what? Describe the slag with words like glassy, opaque, translucent, shotted, stony,
etc. You may find that you made a miscalculation when you mixed the flux, and will be able to correct the problem if you keep track of the results in your notebook.

Retention of Values:

Slags, whether from assaying, or smelting, typically carry some residual values. It is simply the nature of the process. Some slags can carry extraordinarily high values. This is why we will always assay our slags before we decide what to do with them.

Several years back, an assayer watched smelting slags being put in barrels at a small mining operation. The assayer noticed that during the pour, the slag seemed to be too thick. The mine shut down a year later, and our assayer bought the ten barrels of smelting slags for $10,000.00. The mining company was happy to get the ten grand, and the assayer recovered over 500 ounces of gold out of the ten barrels of slags.

A few years further back, a prospector found a pile of old smelting slags out in the desert. About fifty tons, or so. The slags were a brilliant cobalt blue, with a foamy white layer on top of them. The prospector thought the slags were pretty, and unusual, so he carried a five gallon bucket home with him. Later, he had the slags assayed. There were 20 OPT of gold, and 250 OPT in silver. He recovered almost all of the values by crushing the slag and putting it on his leach pad.

The moral of the story is obvious. Check your work.

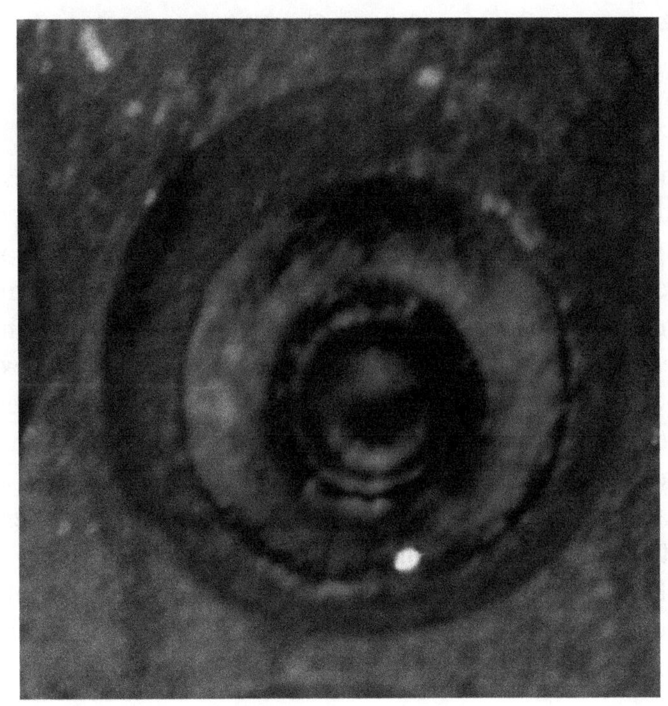

Note unusual color of slag.

Chapter Seven

Crucibles

General:

When smelting, you can be dealing with several different types of crucibles. What you will be using will be determined mostly by the size of your operation. Are you using a small furnace and smelting a lot of small batches, or are you using a tilting furnace with one large crucible? There are a lot of advantages to having a small setup, even if you have a large tilting furnace. You can get your flux recipe down in the smaller crucibles before firing a large batch in a tilting furnace. This way, if something goes wrong, you have a small mess, instead of a large, expensive one.

Remember that the life of a crucible is typically shortened by thermal shock, so each time you use it, inspect it carefully for cracks and erosion.

Check the supplier's list in Appendix C, order a few catalogs, and peruse the crucibles and accessories that are available. You can get a handle on the sizes, styles, and types of crucibles that are available just by studying the catalogs. Try to find a supplier that will sell you what you need, instead of a truckload. Assay crucibles come in cases, and the number in a case will vary from manufacturer to manufacturer. The large crucibles used in tilting furnaces can be purchased one at a time, as can the larger fused silica crucibles.

Assay Crucibles:

Gold and metallic silver can be smelted in assay crucibles, and there are many different brands out there. Some of the better ones are A. P. Green, DFC's regular fire assay crucible, DFC's Colorado crucible, and Liberty crucibles. The most durable fire assay crucible that we've seen is the DFC Colorado crucible. It will stand as many as fourteen firings before becoming dangerous to use. The others mentioned will stand nine or ten firings.

Crucibles come in different sizes, starting at ten grams and going up into the number series, which are quite large in capacity. Most assay labs use either 30 or 40 gram crucibles, meaning that they will hold a 30 or 40 gram charge of ore, and the necessary flux to fire the sample. A 30 gram assay crucible will smelt ten ounces of gold with plenty of room to spare. The inside volume of a 30 gram crucible is slightly larger than one cup.

Assay crucibles, left to right: 20 gram DFC, 30 gram DFC, 30 gram AP Green, and 40 gram AP Green.

Always glaze an assay crucible before you smelt in it. The glaze coats the exposed interior surface, and will prevent metal hanging inside, or weeping through the sides. To glaze a crucible, fill it about one third to one half full of the silver smelting flux (just the flux) and furnace for 30 to 45 minutes, until the flux is molten and fluid.

Swirl the crucible to coat the inside about two thirds of the inside height, and pour the molten flux (slag) into a pouring mold. When the crucible cools, it is ready to be used.

Fused Silica Crucibles:

These crucibles are made of fused silica, and not nearly as sensitive to thermal shock as assay crucibles. The price is correspondingly higher, as well. Fused silica crucibles are not sized by charge, but by letter, then number designations. Size A is smaller than B, and so forth. The numbered sizes are quite a bit larger than the letter sizes. These crucibles can be bought in most any size you can imagine. They are a rose pink in color, easy to identify. The best source to check for fused silica crucibles is DFC. Actually, the fused silica crucibles are an excellent buy, since they are so durable. If you are smelting for hire, you had better factor in the price of the crucible, since you should return it to your customer.

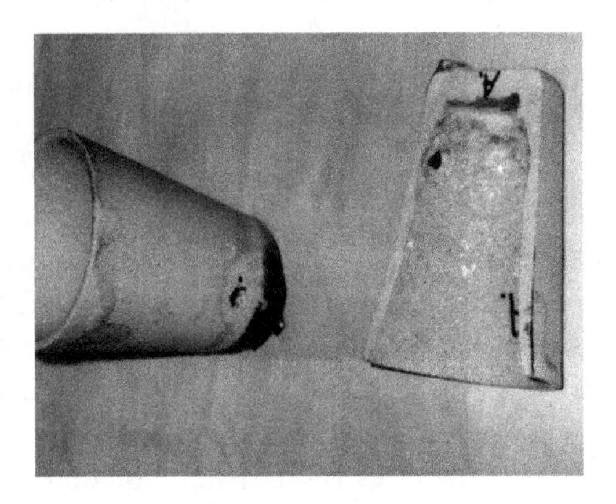

Holes in crucibles are called "burn-throughs", another example of erosion.

Silicon Carbide:

Silicon carbide crucibles are easy to spot, they are black and shiny when new. This crucible is available in most any size you can think of, and are used almost exclusively in tilting furnaces, and large commercial operations. They do erode away on the inside with each successive fusion, and will eventually burn through. If you use this crucible, check for erosion inside after each firing.

This crucible is also unique in the sense that you can buy them with different types of spouts. The "long nose" variety is very popular for tilting furnaces. These crucibles are not as durable as fused silica, and are fairly expensive.

Long nose silicone carbide for Vcella TL-60.

Crucible Management:

If you are using an assay crucible, or a fused silica crucible, get a high temperature crayon from DFC. This crayon will mark the crucibles, whereas a magic marker or ink pen will burn off in the furnace. This way, you can label the crucibles and keep track of how many times you have fired them, what is in them in the furnace, or whatever. Label the crucibles with a simple code like A, B, C and so on. Don't try to put sample names or anything lengthy on the crucible, it will become cumbersome, and eventually very hard to read. You should be writing everything down in your notebook, so make note of what is fired in the crucible each time you use it. When the crucible dies, or becomes unsafe to use, replace it with a new one with the same label.

If you smelt for a friend, or for hire, insist on sending the crucible, slag and metal with the customer. Charge accordingly, which may not make your friend or client happy, but it is better than a dispute later. Make sure they stand and watch everything you do. Save yourself a lot of hassles on down the road.

If you are smelting your own material, you can do about anything you want with the crucibles. Just make sure you know what material was smelted in which crucible. If you have a bad smelt, it can be worth while to fire a "blank" to clean out the crucible for reuse. Just pour it about half full of silver flux, and fire It as usual. This will help clean out any metals that may have hung in the slag in the crucible. **Always "roll" the crucible when you pour.** See the illustrations below. Rolling the crucible will keep drops of molten slag from running down the side of the crucible. The molten slag will stick several crucibles together, and if you have several in the furnace, when you pull one out, it will stick to the one next to it and pull it over inside the furnace. This makes for quite a mess, not to mention the hassle of having to pull everything out of the furnace, scrape the floor, and put down fresh bone ash. If you don't have a layer of bone ash on the furnace floor, the flux will attack and digest the furnace floor, so either way, it can be an expensive mistake. If you don't roll the crucible, the drops of slag will eventually melt down the side of the crucible to the bottom, and star t collecting bone ash on the bottom of the crucible. If this isn't removed after each firing, the crucibles will be setting cockeyed in the furnace, taking up extra space, sticking to other crucibles, and causing spills. So, roll your crucibles.

How To Roll A Crucible

-One-

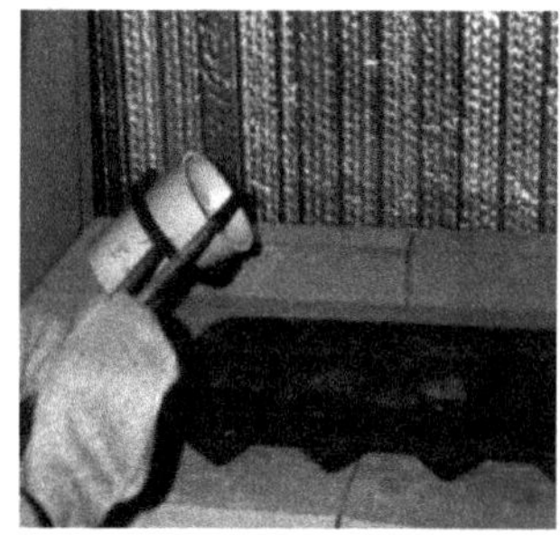

-Two-

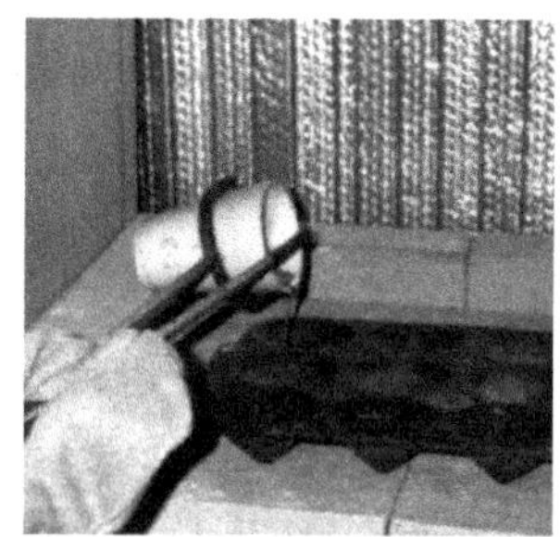

-Three-

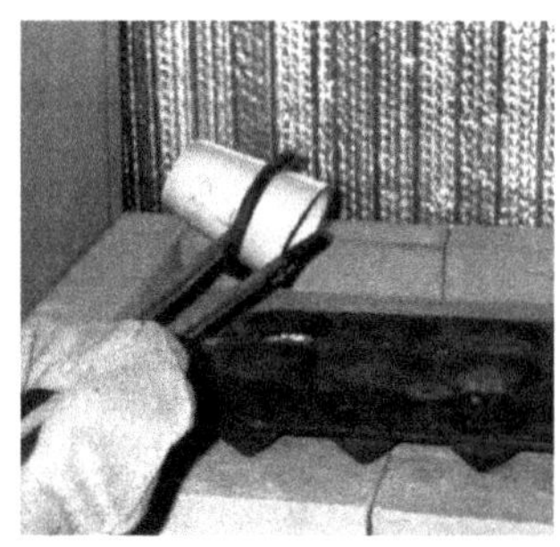

-Four-

-Five-

-Six-

-Seven-

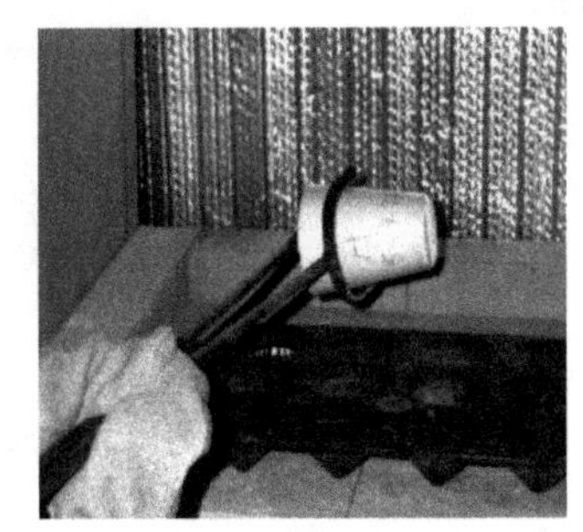
-Eight-

It's really not hard to do. It will save you a lot of grief to learn this from the beginning. See the following explanation of photos:

Photo One- is the start of the pour. The idea is to continue the motion in the same direction, and let that last drop of molten slag run back inside the crucible, instead of down the side. Prior to the pour, the crucible should be swirled and thumped to collect the metal into a homogenous mass. Our person pouring is right handed, and turning the crucible in a clockwise direction.

Photo Two- is the motion of the crucible in the same direction, clockwise.

Photo Three- is the start of the pour, still in a clockwise direction. The slag is visible running into the mold.

Photo Four- is the end of the pour, still maintaining a clockwise motion.

Photo Five- is the crucible continuing in a clockwise motion, to vertical, where the crucible will be shaken up and down, and the last of the slag allowed to run out.

Photo Six- is the crucible in a vertical position, allowing the last of the slag to drain. A novice will now turn the crucible counterclockwise, and allow slag to run down the side. Continue the clockwise motion.

Photo Seven- is the crucible traveling past vertical, in a clockwise motion, as it should be done.

Photo Eight- is the end of the roll, in a clockwise direction. The empty crucible is now turned vertical, set down and allowed to cool.

If you are a lefty, you can do the exact same thing turning the crucible counterclockwise. Try it, you'll like it!

Chapter Eight

Smelting

General:

At this point, you should have an idea how to go about smelting your material. You know how to make the flux, how to mix the flux and the material you intend to smelt, what equipment you will need, and how to go about this in a safe, prudent manner. What you really need to know at this point is what you can smelt, and what you can't.

Remember that we are not trying to wholesale smelt an ore body. If you have this type of operation in mind, be prepared to go to your appropriate State Environmental Agency and secure a discharge, or stack permit. It is a felony offense to do this without one. Fines usually start at $50,000 and go up, way up. This offense also carries some penitentiary time, so think about what you can afford to lose before you start. A discharge permit is required for steam, carbon dioxide, nitrous oxides, or any other atmospheric discharge that is not natural in origin. Think about what automobile manufacturers go through to produce the vehicle you drive.

Also bear in mind that your slags from a production operation will be toxic waste if there is any lead, thallium, arsenic, or other heavy metals present. You will need a permit to create, dispose of, transport, or store toxic waste.

The permits for this type of operation are very expensive, and can take years to secure. Each State has regulations modeled after the Environmental Protection Agency regulations, and the penalty phase of these regulations can be really scary. Do your homework, and know what you are getting into. So much for gloom and doom. Let's move on.

We will start with what you can smelt as presented in this book.

Placer Gold:

Placer gold will probably be one of the easiest things you will smelt. If your placer gold will not pass through a common window screen, you should consider the possibility of selling the gold as nuggets, rather than smelting it down. Of course, it is your gold, and you can do what you want with it.

A button from a dirty con. Way too much black sands. (Photo by H. Geiger)

The main decision here will be which flux to use. Placer gold is always an alloy, so if you use the gold smelting flux you will lose some weight as the impurities are oxidized into the slag. You can "fire polish" the gold to a high degree of fineness by smelting the gold several times, if you wish. If you prefer to keep the metal at the current fineness, and not lose any silver, use the silver smelting flux. You will lose some weight with this flux, since any iron, copper or other base metals present will go to the slag.

The placer gold should be free of black sands, or as clean as you can get it. Follow the instructions given with the flux as far as ratios and all go, mix the gold and flux well, place the flux with the gold in the appropriate crucible, and smelt it. If you have a small furnace, test a small batch of 15 to 30 grams. Weigh the metal carefully for the test, preferably to the hundredth of a gram. If you start with exactly 30.00 grams, and the button (bullion) you pour weighs 27.00 grams, you have a 10% weight loss, and a good idea of the impurities in your placer gold. If you choose to smelt the metal again, weigh the button (bullion) carefully, and note any weight loss. Eventually, you will have a minute weight loss, such as a tenth of a gram or less, and you should not smelt the metal any further, since the weight loss can come from a mechanical error. Usually, two firings will do the job if the placer gold was clean at the start of the process.

Once you have the small tests done, then proceed with the larger batches. If your flux needed adjusting, you should have that down pat before you proceed to the larger batches. Remember that the slag should have no strong affinity to stick to the metal. If it does, add a little of the lime-fluorspar mix to the flux before you do the next batch. Write everything down in your notebook! Save the slags for re-use, and have the slag assayed before you dispose of it.

Concentrate:

Good, high grade concentrate will smelt quite nicely, thank you. Just remember that there must be enough available metal to act as a collector. This is usually 50% or better. If in doubt, dig out your trusty gold pan, and pan a teaspoon full or so of the concentrate. You should be able to see the available metal in the pan, and get a rough idea of the metal available for collection in the smelt. Use the appropriate flux, or smelt first with the silver flux, and smelt again with the gold flux if you wish to take the fineness up higher.

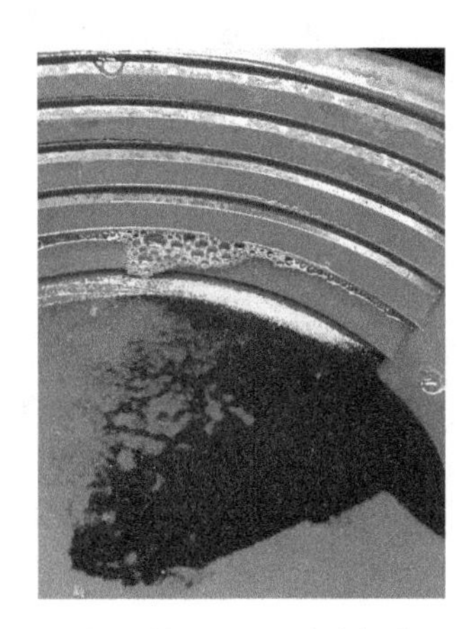

Good Gold, too much black sand. (Photo by H. Geiger)

If you have a preponderance of base metals in the concentrate, you should re-concentrate, or perhaps remove the base metals with an acid pretreatment. You can also roast the concentrate to oxidize the iron compounds. Roast at 1000° C for at least an hour to do this. Never roast the material at any higher temperature, or you will create some real serious iron oxide compounds that are very difficult to work with, and very hard to remove by smelting. A Silmanite roasting dish is great for this, just keep the material in thin layers so that the heat can work all the way through the material. The thicker the layer, the longer you will have to roast.

Bear in mind that roasting concentrate in an electric furnace can be very hard on the heating elements, and that you will evolve toxic vapors. Be sure you have adequate forced ventilation as described in the chapter on safety. It is usually easier to re-concentrate, or run the concentrate on a refining table, such as the a Gemini Table. Miller Dredge also has an inexpensive setup that will also clean up concentrate very well.

Scrap Jewelry:

Not a good idea to smelt this directly. Remove all the stones first, or you will destroy them. One thing you can do is take the jewelry, a large amount of scrap silver, and use the smelt as an inquarting process, which will allow the separation of the precious metals as the assayers do it. This is usually done by parting the silver with dilute nitric acid, then smelting the gold that is left. The silver is normally recovered as a chloride, converted back to metallic silver by cementation, and smelted separately with a different flux.

If you do process jewelry, make sure that each piece has an obvious karat marking, so that you aren't trying to recover the gold in plated items. Plated jewelry is not economic to work with at the current gold prices.

Precipitates:

This word can mean anything to anybody, so make sure you know exactly what these "precipitates" really are. Normally, the word indicates that the material would be precious metals precipitated from a chemical solution, such as an aqua regia digestion. If that is the case, and the precipitates are gold, they will smelt quite nicely. Be careful when handling the dried powder, you can lose a substantial portion as dust if you are not careful. Pour slowly. This material is normally a dense brown powder, and should be dried thoroughly before you mix it with the flux. Usually it will be of a very high fineness to begin with, so one pass is normally all that is necessary to produce marketable bullion. As with anything you are working with, record all weights in your notebook, and if possible, do a small test smelt first. Make sure that the precipitates were rinsed with de-ionized water at least three times, or the residual acid will erode your crucibles faster than normal.

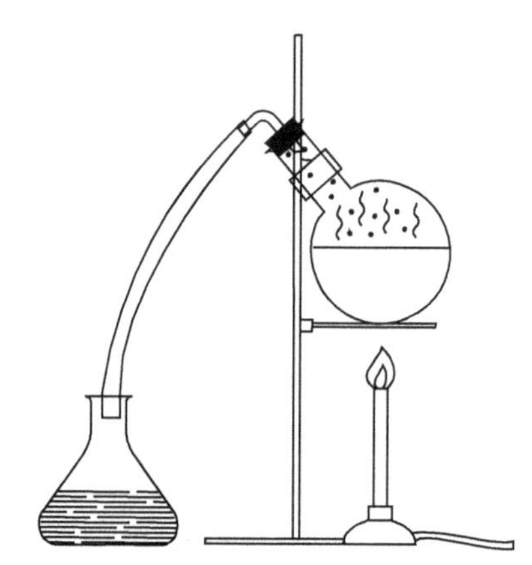

Wet chemical pre-treatment may be necessary for some materials.

Precipitates #2:

This again, can be about anything, but is generally the material removed from plates in an electrolytic cell, as in electro winning. If someone brings you a bunch of steel wool that gold has been plated on from solution, have them take it back, and digest the steel wool with sulfuric acid. The iron in the smelt will be preponderant from the steel wool, and your smelt will be marginally successful, if at all. Again, make sure the product you have is rinsed well, and absolutely dry before you mix it with the flux. Be very careful, this stuff is usually a very fine purplish powder, and dusting losses can be serious. So can contamination.

The other type of material you will probably see are the scrapings off of stainless steel plates in an electrolytic cell. If it was done right, this material will be fairly coarse, and easy to handle. If it is a very fine black powder, advise your client, or if it's your operation, add a little lead acetate to the solution during the electro winning operation. This will make the "precipitate" a lot coarser and easier to work with. It will also make removing the material from the plates a lot easier, since the lead acetate will cause the deposit on the plates to be "fluffy", coarse, and have a tendency not to stick. To many volts can also cause the material to weld to the plates.

In any case, a small test smelt is a must. This type of material will contain large amounts of base metals and can be ver y difficult to work with. If it doesn't smelt well, upgrade the material with an acid pretreatment, and try again.

Silver Chloride:

Yes, it can be done, but is a pain in the butt to work with. Never use an assay crucible to smelt silver chloride. Most silver chloride fluxes do not work.

Normally, the silver chloride is reverted to metallic silver by cementation, usually with scrap iron or zinc powder in dilute sulfuric acid, rinsed well, dried and then smelted.

Silver chloride can be direct smelted by placing it in a fused silica crucible, mixed with flux, of course, and heated in stages to drive the chloride portion off as chlorine gas. Usually, the temperature is taken to 600° F, and held for an hour, then the temperature is raised to 800°F and held for an hour, then taken to smelting temperature and held for another hour after reaching temperature. The slags are smelted again with the addition of potassium nitrate for two hours at temperature to determine if any silver was hung in the slags.

Never use assay crucibles for silver chloride. Note erosion and perforation.

It is a lot simpler just to cement the silver, and smelt. It is always a good idea to smelt the slags again with the potassium nitrate just to check your work. Another good idea when working with silver is to pour to a graphite mold. Silver, at the temperatures you will be working at will have a tendency to "sprout". This is caused by the cooling process. Basically, what happens is that the silver will cool on the outside, and still be molten inside. The mol ten metal will squirt out of the bar, sometimes with spectacular results. A graphite mold allows the metal to dissipate the heat at a more uniform rate, making sprouting less likely. Always give silver extra time to cool before you break the slag away. The same rule applies to any alloy that has a high silver content.

Silver (Metallic):

This will smelt very well. Keep in mind all the information for silver chloride as you go, and don't use an assay crucible for smelting silver chloride.

Some people smelt sterling silver, thinking that they can slag off (oxidize) the copper used in this alloy. It doesn't work real well, it is much better to chemically separate the base metals first. Keep in mind the economics of the situation. It is usually a lot cheaper to do a chemical separation than repeated smelting, and the end result will be much better if you have chemically separated the base metals first. Also make sure that what you are working with really is sterling silver, not silver plate. Normally, the silver plate is over a lead tin alloy that can really cause problems for you.

Some items marked "German Silver" are not silver at all. So do your homework, and know what you are dealing with before you expend a lot of time and effort.

Silver concentrates are around, and will smelt, assuming there is enough free metal to act as a collector. Normally, these concentrates are chemically separated to recover the gold, then reverted to metallic silver for smelting. If you happen to have a lot of high grade silver concentrate to smelt that has no gold present, you can smelt to your heart's content. If you have some gold present, and are selling your bullion to a refiner such as Johnson-Matthey or Englehardt, the gold will be treated as an impurity, and you won't be paid for it. In fact, you can even incur penalties for the gold, as an impurity. Something to think about. It really doesn't take a lot of gold to make the chemical separation worthwhile.

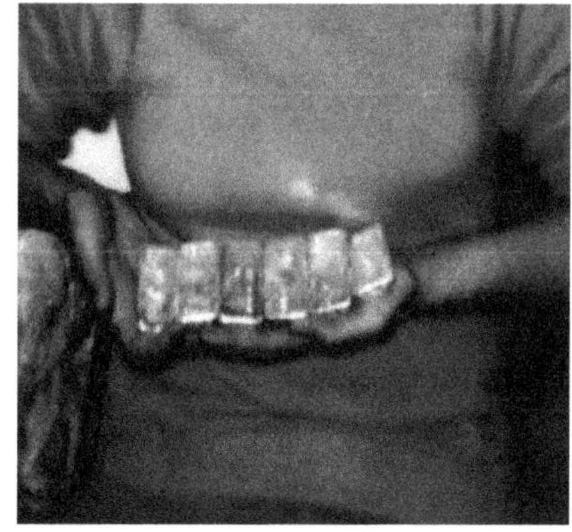

Silver produced using the methods and flux in this book.

Silver is unusual in the sense that there are actually deposits out there, especially in Mexico, where the grade is so high that the ore can be crushed, pulverized, dried and directly smelted. There isn't a lot of this around the United States or Canada these days. Most of that grade was mined a long time ago, but will pop up now and then.

Carbon Ash:

This is produced by ashing activated charcoal, or carbon used in the mining industry. Typically, the carbon is produced from coconut shells either in Sri Lanka, or the Phillippines. There are different grades of carbon available, and some crafty operators have figured out that it is cheaper to use a finer, high activity carbon than what the major mining companies use, load it to the maximum, and ash the carbon. The ash is then smelted, usually after an acid pretreatment to remove the excess iron. The mines use a coarse, medium activity carbon and only allow the carbon to load to certain levels, usually about 400 - 600 OPT. It is almost impossible to strip higher values from loaded carbon so that it can be reused. By using a finer grade of carbon at higher activity levels, it can be loaded to several thousand OPT, and then ashed. The ash will carry enormous values, since the volume is reduced, hence, upgrading the assay. A ton of carbon will yield about two hundred pounds of ash, which contains all oxide material from the heat generated in the ashing process.

Once an acid pretreatment has removed the excess iron, the carbon ash will smelt like a dream using the appropriate flux. If you pretreat high silver ash with acid, be prepared to chemically recover the silver.

Our smart operator has not had the capital expenditures to build a stripping plant, and doesn't have to work with a lot of really nasty chemicals, or heat and pressure. Not to mention the extra manpower required to run a strip operation. So, you may smelt some carbon ash. Be sure to test smelt small batches first. Carbon ash will have visible metallic gold, or silver, if that is what the operator was after. Just use the appropriate flux, and smelt away.

Amalgam:

Never smelt amalgam. The mercury that will vaporize is very toxic in small quantities, and can be lethal in large quantities. A much better method is to digest the mercury with dilute nitric acid, rinse the remaining material well, dry it, and smelt it. If you have to filter the material after the mercury is digested, ash the filter paper and include it in the smelt. You can recover the silver (if any) from the nitric acid, as well as the mercury. Scrap copper or aluminum is normally used to recover the mercury from solution. Remember that mercury will start evolving toxic vapors at room temperature, and always keep it in a non-breakable container, covered with water or light oil.

Mercury fumes will kill, sooner or later.

Sampling:

There will come a time that you will have to sample the metal that you smelt for an assay. The absolute best way is to use the vacuum "pin tubes" that were developed for this purpose. Pin tubes are glass tubes that have a vacuum in them, and have a blister on the end. You shove the pin tube in the molten metal, the blister melts, and the vacuum pulls a couple of inches of molten metal into the glass tube. After cooling, the glass is broken away from the pin with gentle taps of a hammer, and you have your homogenous sample. The reason the sample is homogenous is the metal in the pour is in a state of molten flux. After the metal cools and sets up solid, the heavier elements such as platinum, if present, will settle to the bottom. The lighter elements, if present will tend to be towards the top of the bullion. In other words, the metals present segregate themselves according to specific gravity as the metal cools.

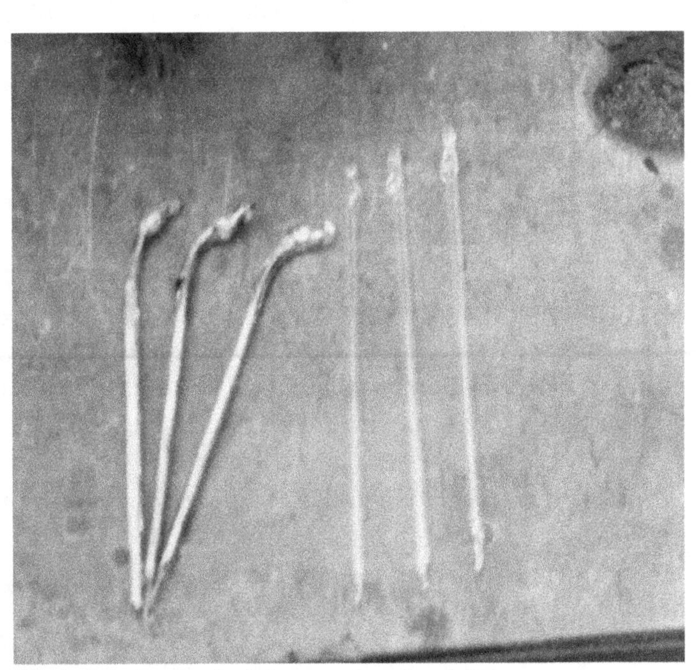
Used pin tubes on the left, unused on the right.

As second best, the metal can be drilled with an eighth inch drill bit for a sample. Usually, three holes are drilled all the way through the bar, for obvious reasons. The idea here is to get a sample that is representative of the whole. The pin tubes are much more representative than the drill cuttings. If you don't have pin tubes, well, drill the metal, and hope for the best.

Pin tubes are available at assay supply houses such as Legend, Inc. Check the suppliers Appendix. Everyone has a drill and bits.

What You Can't Smelt:

Dirty concentrate. Low grade head ore. Or high grade head ore. Any material that does not have the necessary metal to collect in the smelt. What we have here is a process that will upgrade the right material to a marketable product. Try to avoid plated jewelry and such items, since you will still have to remove the base metals after you have smelted the jewelry. Keep this process in perspective and understand the purpose behind it. Don't try to smelt high grade sulfides, either. The sulfur will reduce the base metal contaminants, and you'll spend a lot of time and money dealing with the separation problem you have created. Pre-treat the sulfides chemically, or with heat, concentrate, and then smelt.

Odds and Ends:

Sooner or later you will run into other people who smelt precious metals, and learn new information. One of the things you will encounter will be the use of a "crucible wash". A lot of old timers used common, un-iodized table salt in their smelt to "wash out" the crucible. The idea is that the salt, when molten, will float on top of the smelt, and being very liquid when molten, will cause all the slag, metal, etc. to flow out before the salt during the pour. It does work, however some schools of thought do not agree. The argument is that the gaseous chlorine that is produced at temperature will cause "dusting loss" of the gold during the smelt. This is one that you will have to decide for yourself. It is not necessary to use the salt with the flux recipes you have in this book, however, if you decide to try salt, remember that a little goes a long way. An excess of sodium will be visible as a clear, watery liquid when molten, and will ride on top of the smelt. When you pour, the sodium will pour last, and cool to a white or off white cap on the slag. Remember that the salt, sodium chloride, is no more. The chloride portion of the compound has gassed off as chlorine, and only sodium remains. Other schools of thought feel that the salt will decrease the life of the crucible by contributing to erosion. So, it is up to you.

Never, ever experiment with your customer's material. If you run a small test smelt, that's fine, just keep everything together. If you are altering or adjusting fluxes, or just plain experimenting, use your own material for the experiment. If something goes wrong, you will have an upset client on your hands. Remember that you are accountable for the material you are working with, and invariably, the material is worth a lot more when you can't produce the amount of metal the client feels should be recovered. Most people you deal with will have an assay on the material, and have a fair idea what you should be able to recover.

Chapter Nine

Shotting

This information was originally going to be an article for the International California Mining Journal. Having had inquiries into the process, it seemed much more useful to include the information in this book.

General:

The process is straightforward, and the tools required are common. Obviously, the first tool would be a heat source capable of creating enough heat to melt the metal one is working with. Since most metals are shotted in small quantities, a furnace that can hold several 40 gram assay crucibles will work nicely. It is wise to shot a small amount and check the results prior to moving on to larger batches. Typically, a two or three ounce pour is shotted, and if the desired results are acceptable, than larger quantities can be processed.

Pouring the molten metal. Notice the tongs hold the crucible for a safe pour.

The gear required is that typical of any assaying or smelting operation. Heat resistant gloves or mittens, face shield, respirator, adequate forced ventilation, and long handled crucible tongs with a safety bar to hold the crucible in place so that it can be inverted. Great care must be used not to drop the entire crucible into the water.

The size of the chunks of metal is predicated on the speed of the pour. For refining purposes, a faster pour produces smaller chunks, where a slower pour will produce larger artificial nuggets. The photos that accompany this document show a slow pour and a large pour.

Shotted metal. A slower pour into the water produces larger pieces.

Shotting metal can be a useful technique for the prospector, or miner that would like to increase the profit on metal that is produced, or in the lab, shotting can be useful as a refining technique. Artificial nuggets can easily produced if one has the equipment to smelt. Artificial nuggets must be represented as what they are, not natural nuggets. A quick search on the Internet can provide sources of jewelry making supplies to mount, chain or set the nuggets.

Digesting Or Refining Silver:

From a lab standpoint, gold can be refined with a massive silver inquart. The inquartation process is common in the fire assay. Silver is added to the fire assay in the amount necessary for the assay bead to part, usually a ratio of four parts silver to one part gold. In some cases, the ratio is raised to prevent the loss of gold in the fire assay from the presence Tellurides or some rare earth elements. After cupellation, the assay bead is parted (digested) in dilute nitric acid, HNO_3. Once the bead has digested, it will appear as a small, fluffy black speck. The bead is annealed at a lower temperature than the assay, the black fluffy nitrates from the digestion are expelled, and the gold is visible as a dark yellow or brown speck. After annealing, the bead is weighed, the assay is calculated and reported. The silver solution from the digestion of the bead is saved, the silver to be recovered and recycled at a later date in most labs.

Cementing Silver. The skillets provide the necessary iron. Zinc powder is a lot faster.

The silver nitrate solution is treated with saturated salt water or hydrochloric acid (HCL) to produce a dense silver chloride. Exposure to sunlight will cause the silver chloride to turn any color from pink to black, according to the purity, among other things. If you suspect lead chloride, simply rinse the silver chloride with very hot boiling water several times, or boil the mixture on a hotplate, and pour the mixture through a filter. When the water cools, the lead will re-precipitate out of the water. Mercury is not an issue at the point due to the temperature used in a smelt, and will have vaporized.

Cement your clean silver chloride with powdered zinc in dilute sulfuric acid, (H_2SO_4) or with heavy white syrup in dilute sodium hydroxide, (NaOH). The cement silver will smelt to metallic silver with any silver flux, and will not bleed through a crucible like silver chloride. Silver chloride can be direct smelted after drying in an expensive fused silica crucible, if you have one.

Suppose you have a few ounces of 50 or 60 fine gold, and want to refine it. You would want to have a good idea how much gold you have so that the appropriate inquart could be added. As an example, if you have two ounces of a gold alloy that is 50 fine (50 %) gold, you would add eight ounces of silver as an inquart. Note the four to one ratio. If the alloy is higher in gold, it is always a good idea to add extra silver to ensure a complete part. In the previous example, 50 fine gold is a pale, light yellow. If the gold were darker, say 85 fine, it would be a good idea to add an extra ounce of silver to ensure the parting process. If you are an assayer, cupellation of a small amount of the sample can give you a ratio of the gold in the alloy. If you have had a bullion analysis done on the alloy, you have the information you need.

If you are seeking the best part ever, and want the highest purity of gold, it helps to add a small amount of copper. Clean scrap copper works well. In the previous example, a quarter of an ounce should do the trick. The copper is considered waste, and will follow the spent acid after digestion and silver recovery.

If you are making artificial nuggets, you won't be digesting the shotted metal. The trick to making the nuggets is to remember that a fast, quick pour into the water will produce larger chunks than a slow, thin stream being poured into the water. Don't just dump the crucible contents in the water wholesale, steam will blowback.

Colored Gold:

The color of artificial nuggets can be anywhere from pale yellow to pink, dark pink, red, green, blue, purple and even white, depending on the alloy used to produce them. Wikipedia has an excellent article on this subject, search "colored gold". Green gold is made with toxic cadmium and should be avoided. Note that cadmium is skin absorbable, and should not be used in any application that touches the skin. Also remember that the jewelry alloys are measured in karats as opposed to fineness. Each karat indicates 1/24th of the whole. So 12 karat gold is 50 fine, 24 karat gold is pure gold. The Wikipedia article has the ratios of metals used to create the individual colors.

A smaller quantity of metal would be shotted to determine the color of your alloy, and if the results are satisfactory, a larger pour could be done.

Larger scale pours are done for refining purposes, and can be done in Troy pounds, or multiple ounces, as previously indicated.

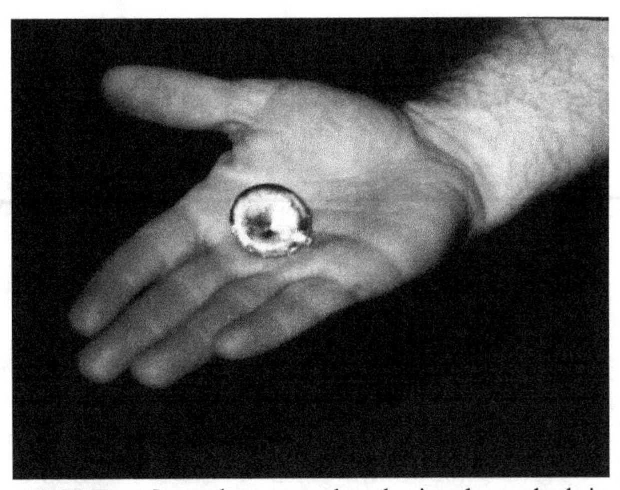

A 3.87 Troy Ounce button produced using the methods in this book.

Chapter Ten

Useful Information

General:

First and foremost, understand the word "fineness". What the word means, precisely, is the proportion of pure precious metal in an alloy, often expressed in parts per thousand. In other words, if an alloy of precious metal is 913 fine, it is 91.3% gold, or silver, or whatever the preponderant precious metal is. Base metal content is not expressed in fineness, only precious metal content is expressed as fineness. Absolutely pure metal, say gold, would be 1000 fine, or 100.0% gold. This is rare, normally precious metals are .999 fine when refined to the most reasonable purity that is cost effective. If you follow the reasoning here, placer gold that is .813 fine is 81.3% gold.

Fineness is determined by bullion analysis, usually of pin tube samples of the metal taken at the time of the smelt, when the metal is poured and still in a state of molten flux. The pin samples are submitted to an assay lab, where the analysis is performed, usually in triplicate.

Keep your trusty calculator at hand. Use your notebook.

Another word that you will encounter is "hallmarking" or "hallmarked". What this means is that an official mark or stamp indicating a standard of purity, used in marking gold and silver articles has been affixed to, or imprinted on the metal. This is done by the large refining companies throughout the world, *who are licensed to do so by their respective governments.* At one time, years back, it meant that the metal had been assayed by the Goldsmiths' Company of London, and the corresponding purity of the metal had been stamped into the metal.

If you were to go to Englehardt, or any other major refiner, and purchase hallmarked metal at .999 fine, re-smelt the metal, and try to sell it back to the refiner, you will lose about two percent. That is what they will penalize you to determine the purity of the metal. That also explains why coinage is valuable to people who accumulate gold and silver. Gold coins are typically around .900 fine, and the coinage, by virtue of being stamped, tells you immediately what you are dealing with and what it's worth. The added value to collectors usually makes the coinage worth more than the gold that is present in the coin. If gold is at $400.00 an ounce, and you have a double eagle that weighs one ounce, .900 fine, it is worth about $360.00 for the gold. Try buying a double eagle for $360.00. You will find the price a lot closer to $1200.00 or higher, since the coin is a collector's item.

By the same token, an ounce of placer gold that is .800 fine would be worth about $320.00 with gold at $400.00 per ounce. The problem arises if it is a one ounce nugget, which is worth more than an ounce of dust. The other thought to keep in mind, if you're buying gold, who will pay for the assay? Buyer or seller? A bullion assay can set you back several hundred dollars, depending on where it is done. If you are buying any kind of quantity at all, you had better take your own sample, and have your own bullion assay done. Maybe there are some brass shavings mixed in, or whatever.

It has happened. You may have to pay for the material you send in for analysis, but the recovered gold will be returned to you by any scrupulous assayer. Be sure to specify that the gold is to be returned.

Another term that gets a little wild is "karat". Again, this is a unit for measuring the fineness of gold, pure gold being 24 karat. The catch here is that the jewelry industry is allowed a half karat either way. So, if you buy a gold stickpin that is stamped 12 karat, it could be anywhere from 11 ½ karat to 12 ½ karat, a spread of just over 4%. You don't suppose it could be on the high side, do you? Better plan on a loss of 2%. It will be 11 ½ karat, most likely. If you are smelting scrap jewelry, you should be aware of this, and plan on the shortage.

Make More Money:

In the next Chapter, you will see a lot of information about selling your metal. Something that you should keep in mind as you read this information, is that you can make a lot more than spot price on your metal, and eliminate most buyer discounts by making your metal into a more marketable, desirable product, such as jewelry. All trade publications have information on the accessories for jewelry, and you should seriously consider this.

Coinage can be worth a lot more than face value.

How about artificial nuggets? Put a small amount of metal in a crucible (a small one) with a little borax, melt the metal, and pour it on wet sawdust. Or pour molten gold into half of an English walnut shell. You will have an artificial nugget that can have a chain attached to it, making it a pendant. Remember to create an alloy first, don't use pure gold or silver. Add silver and a small percentage of copper to the gold, and add copper to the silver to create the alloy. The nuggets you make are worth at least twice what the gold in the alloy is worth.

Another angle is to color your alloy with different metals, then make the artificial nuggets. A jeweler will set stones for you, or you can learn to do this yourself.

If you can afford it, think about buying a wire machine. You create an alloy, say 12 karat, cast it into the appropriate shape for the wire machine, and let the machine extrude the gold (or silver) alloy into a given wire size. Since all jewelers use the wire for soldering and such, and in the manufacture and repair of jewelry, it is a very popular, expensive item. Wire machines are available in the jewelry industry, you'll have to get trade publications and track down suppliers, but it could be well worth the time and trouble.

Industrial Cupellation:

Sooner or later, you will encounter this process, and would probably wonder why the subject wasn't covered in this book. This subject is not included with the smelting information because it is contrary to the purpose outlined in this book.

Industrial cupellation is an oversized fire assay. A flux containing litharge (lead monoxide) or metallic lead, soda ash, silica, and borax glass is mixed with the material, and it is heated and poured, just like a fire assay. The lead is the collector of the precious metals. The lead is removed with the process of cupellation. The lead from the fusion is cleaned of slag after cooling, and placed in a refractory dish made of bone ash, magnesite, Portland cement, or a combination of these materials. The cupel has a hemispherical depression in the top to contain the molten lead, and focus the heat, which superheats the lead. The cupel will absorb about 90% of the lead, the remaining 10% or so goes into the atmosphere as an oxide. Ventilation is very important.

New cupels, with lead buttons in them. Cupels are normally preheated.

The precious metals are left in the cupel as a large, rounded blob of metal called Dore'. The Dore' is then refined by the usual means, or sent to a commercial refiner. Large operators that use this process generally make their own cupels. The cupels are enormous compared to those assayers use, which are rarely over two inches in diameter. See the illustrations.

The major drawback to this process is the lead, and lead contamination of the surrounding area. The technicians routinely have elevated blood/lead tests. Lead is toxic to the human body, and is inhaled and ingested. Clothing from technicians should be washed separate from clothing worn by children. The lead waste (slags, crucibles, flue dust, and cupels) created in the process will have to be permitted for disposal, since they are toxic. This isn't cheap, and hello, EPA.

A cupel after absorbing lead. Note the Dore' bead in the center.

There are some major health hazards involved in this process. *If you can, avoid it at all costs.* If you can't, get monthly blood tests (heavy metals) to monitor your health. Some assay labs have switched to bismuth oxide as a substitute for lead, however, the results have been marginal. Industrial cupellation will work on lower grades of material than you would normally smelt, as well as on the higher grades. You will have to decide for yourself about using this method. Keep the operation away from your children and pets if you do.

Chapter Eleven

Upgrading Your Concentrate

Observations

You have a 450 OPT gold concentrate. You smelt a hundred pounds of the concentrate, and pour a 50 Lb iron slug. No visible gold. How can that be, you ask. The simple truth is that a 450 OPT assay is good, but way too low for the smelting process. Somewhere in that 50 Lb slug is a BB of gold, or maybe several BB's. All you have done is create a separation problem for yourself. It will be an economic loss to recover the gold from the iron. You can re-smelt until the cows come home, but there is too much iron to overcome. There is a limit to what the flux can oxidize into the slag.

If you had a 14,500 OPT gold assay, this would be more appropriate, and there would be enough gold to collect in a smelt. You will note throughout this text that it is recommended that your concentrate, fines or whatever, *should be at least 50% gold.* That is 50,000 parts per million. That grade of concentrate should never be sent to an assay lab. The contamination risk would be way too high for the lab. The concentrate could be digested and analyzed on an ICP, but would have to be diluted out many times to get in the range of the ICP.

Common sense would dictate that a simple gold pan would resolve the issue. A teaspoon of the concentrate could be swirled around the pan, and you would be able to see if the material was half gold, or more. Placer miners are used to looking at placer fines in their cleanup, and rarely have this problem. Placer miners will also have a method of cleaning their fines to acceptable smelting limits.

Good gold, too much black sand. (Photo by H. Geiger)

This problem, low values for a smelt, are more common in a hard rock environment. The ore is mined, crushed, pulverized and roughly concentrated. The problem can resolved by many different methods, and the solution will have to be based on the scale of the operation. How many tons per day are processed? How much rough concentrate is produced? Base the equipment capacity on the production of concentrate.

Options

As a case in point, a few years back, some friends had obtained the coarse reject from a milling operation in Idaho. The operator had never ran screen fractions on the ore, had processed for the coarse gold, and left several ounces per ton in the oversize. The friends were given the material on the condition it was moved off the mill site. The material was moved to another property, where a quick analysis was done. A Bico disk pulverizer of the type normally used in assay labs was located, and a Gold Cube was purchased. The pulverizer would run about 200 lbs per hour. There were three people involved, so on days off and weekends they processed the material, recovering 99% percent of the gold. The equipment was dirt cheap compared to the recovery.

Sometimes, it is best to think out of the box. Fortunately, these folks were experienced in mining, and understood what was there, and the value of a simple fire assay. So, when you see some of this "mini" equipment, don't turn your nose up. If the small equipment will duplicate the mill or recovery circuit you have in mind, try it.

If you have less time, and more material, think about a lab sized concentrating table. There are many types of these out there, the best being Wilfley, Deister, and the top of the line lab table, the Gemini. Tra-Lite tables are no longer made, but were nice tables with a very clean cut.

A concentrating table at work.

Chapter Twelve

Selling Your Gold

General:

This is the best part. If you've done your homework, and paid attention, you will be casting a very pure product. If you are smelting the same material all the time, it would be wise to have a bullion analysis done. This way, you will know the fineness of the metal, and will not misrepresent it. Some alloys appear to be quite a bit purer than they really are due to copper that is alloyed with the gold. You can always run a small batch with the silver flux to see if any copper shows in the slag.

Jewelers are good people to know when you need a second opinion. Sometimes they will have touchstones and can give you a very accurate estimate of the fineness of your gold. There are no simple tests for silver. Spot tests with acids generally will tell you the metal is silver, but you already knew that. This is a "qualitative" analysis, rather than a "quantitative" analysis. There are a lot of people out there that think they have a magic eyeball, and can look at your metal, and tell you the fineness. Not true. Have the assay done.

Ideally, you will have a small furnace to test smelt in, and you can take it one step further, and learn to do the analysis yourself. You must have an accurate bead balance to do this. Don't try measuring the bead with an optical device and calculating the weight. It won't work. Neither will calculating the "squat factor". If you're going to assay, well, do it right. If you are in a production situation, it will be worth your time to learn the business. If it is an occasional thing, maybe it would be best to send the metal out for an assay.

Security:

In a day and age where convenience store clerks are killed for $35.00 or less, this subject shouldn't have to come up. But with the so-called "mystique" that surrounds gold, well, we'd better get it out there. Understand one thing. Your worst enemy is your mouth. And yes, it's pretty impressive when you accomplish something of this nature, but let's not pay for it with your life. Be very careful who you talk to, and remember that your very best pal has his own very best pal, and he will share your secret. The word is then out. The idea here is not to induce paranoia, it is to instill a healthy sense of caution. In the western United States, it isn't quite as bad as other places. A lot of people carry around a little "dust", or a nugget or two. But then again, they don't wear it on there chest, either. You really need to think this issue through. It is best not to advertise what you have.

Never use a document/fire safe for valuables. Use a floor safe.

It was mentioned earlier that a floor safe was the rule in secure storage. Don't run out and buy a fire safe that rolls around on wheels. Some enterprising soul will roll it right out of sight, sooner or later. The big thing for years was to break in, and hook a cable from a winch or wrecker to the safe, winch the safe through the wall, and drive away. It does work, and takes very little time to do this. A floor safe set in a yard of reinforced concrete is a whole different story. If you contact a reputable dealer, he can provide you with the statistics of floor safes as opposed to a safe that sits on the floor. There are instances, very common, where a couple of burglars have worked on floor safes for a whole weekend without gaining entry. Most of the citizens of this country are very security conscious, and alarm systems are very reasonable these days. So think about securing your assets. And remember that none of this is worth getting hurt or killed over.

Dealing With the IRS:

Actually, you probably won't have to worry about this, if you have a few smarts. To the IRS, gold is simply a commodity. You are legally required to report any cash you receive from the sale of your gold as income. Just like any other income. What gets people in trouble is trying to avoid paying the tax. If you are running your business like a business, you can deduct the cost of everything you do as business expense. You can depreciate your equipment, write off the utility and phone bills, and about everything else. It really makes a lot of sense to get in the system and be professional about this.

If you are sneaking around trying to peddle a few ounces of gold, sooner or later, an ex will get you. That's an ex-wife, ex-friend, ex-partner, and so on. The IRS won't get real interested unless there is serious cash involved, or a good Tax Evasion charge a possibility. Then they will send someone to buy your gold. Or they'll have your ex do it, and pay them a reward. Your butt will be in the cooler, and it isn't worth it. Have your cake and eat it, too. Be legit. Work from within the system. Be professional. You will go a lot further than a jail cell, or bankruptcy from having to defend yourself in court.

Don't mess...with the IRS!

Marketing:

Strangely enough, this is the easy part. Your local jeweler will buy gold from you, and a lot of doctors and lawyers will buy gold as a hedge against inflation. A lot of people these days will buy small quantities of gold to put back. These are the intelligent people that take a good look at the political economics of this country, and realize where we are headed. You will, no doubt, be smart enough to put back some of your own handiwork, as well.
For years you could sell gold and silver directly to the U. S. Mints. Alas, those days are gone, so you'll probably have to develop your own marketing.

Never pour huge bars. If you've got that kind of metal available, you've been dealing with the world wide refiners, such as Englehardt or Johnson-Matthey, and know how the game is played. If you have a small operation, keep the ingots or bars small. You will sell them twice as fast. It's a lot easier for someone who is working for a living and raising a family to come up with $300 or $400 as opposed to thousands. If you do go to the commercial refiners, be prepared for a shock. They usually want at least 1,000 Troy ounces of silver, and 10 Troy ounces or more of gold. They will assay your metal, and charge you for it at settlement, usually five or ten working days. They will also charge you a refiner's fee that can go from four to six percent, or more. You will probably have the choice of a cash (check) settlement, or you can have your metal (most of it) back as hallmarked metal. So, think about that approach, if you have a lot of metal to sell, as in Troy pounds.

If someone offers you cash, explain that you will be reporting the transaction, and suggest that they do, too. Give them a break, let them know where they stand at the onset. If you accepted the cash, and your customer was an informant or an IRS agent, well, you're history! If you do take the cash, make sure to fill out a receipt with the buyer's name, address, etc., and offer the buyer a copy. At least you will have covered yourself. If your buyer gives you phony information, well, at least it isn't on your head.

Another encounter you will have, sooner or later, is with a bad guy who wants you to smelt his gold or silver. This person will have an fairly large amount of metal in the form of jewelry, table service, or whatever. If the material is obviously not scrap, all worn out, or damaged, well, look out! Some of these guys have hallmarked metal they will want you to smelt. Does that make a lot of sense? Knowing the loss they will take by smelting doesn't make sense at all. They are most likely trying to disguise stolen merchandise. Don't play the game, it will take you to the road to ruin. To possess or alter stolen merchandise is a felony offense in most states. To purchase stolen merchandise is also a felony offense. So remember the old adage, "Good Deals Usually Aren't" and act accordingly.

What's It Worth?:

Well, this isn't too hard to figure out. Find out the spot price for gold on the date of sale. This is usually published every business day in the major newspapers, in the financial section. Suppose it is exactly $1200.00 per ounce. You should know the fineness of your metal, suppose it is .800 fine, or 80.0% gold. Find 80.0% of $1200.00 with your handy calculator, it should be $960.00. So theoretically, your gold is worth $960.00 per ounce *on **that business day***. Since the price of gold does fluctuate from day to day, you'll have to pay attention to spot.

Silver is the same process, check spot prices and get the quote for the day. Suppose it is $15.00 per Troy ounce, and you have 50 Troy ounces that are .965 fine. Multiply $15.00 by .965, and you will get $14.48 per Troy ounce..Your silver is worth $724.00.

In reality, a lot of buyers will expect a small discount of two to five percent on the metal, especially if it isn't hallmarked. The buyers feel that this offsets the risk of buying metal of an unknown fineness. You could show them an assay report for the bullion analysis, which will minimize the discount, but not eliminate it. And let's face it, the commercial refiners will charge you more than that. It's your metal, and it's

You can figure it out!

your decision. As time goes by, and your credibility is established, the discount should get smaller with your regular buyers. Haggle! That's part of the fun!

Don't expect to get paid for the small amount of silver in a gold bar. It is considered an impurity, and would cost more to remove than the silver is worth. If your bar weighs 75 Lbs., and is half silver, well, that's a different story. You should be aware of these things. If you ask to be paid for the silver in a 10 Troy ounce gold bar that is .965 fine, your buyer is going to laugh you out of town, or get really upset. Think about it. You have ten troy ounces, or 311.035 grams of metal. (One Troy ounce is 31.1035 gram s) So, 300.149 grams of that metal is gold. The other 10.886 grams are ***mostly*** silver. There is no way that you can extract those 10.886 grams for $38.58, which is what the silver would be worth at $15.00 per Troy ounce. That's assuming the 10.886 grams of metal is pure silver, by the way. Don't waste your breath haggling on this one. The gold is worth $11,832.00 (9.65 Troy ounces) at a spot of $1200.00Per Troy ounce. Grab it and run, forget the silver, it isn't even worth mentioning in this case.

Figuring the value of your gold is simple if you stay with grams, as in the previous exercise. Remember that 31.1035 grams equal one Troy ounce. You just have one number you have to remember. To know what your metal is worth, all you have to have is the fineness, and spot price out of the newspaper. Try a few practice runs, and you will see how easy it is.

Gold in this form is easy to sell. (Photo by G Vincent)

Good Luck! Enjoy!

A good online resource for mining terminology is www.infomine.com/dictionary/.

Acid rain- Precipitation containing acid-forming chemicals, chiefly industrial pollutants, that have been released into the atmosphere and combined with water vapor: ecologically harmful.
Acid- A substance having a pH value of less than 7. See pH.
Activated Charcoal- (or carbon) a form of carbon having very fine pores, used chiefly for adsorbing gases or solutes, as in various filter systems for purification, deodorization, and decolorization. Generally made from coconut shell by burning in a reducing atmosphere.
Adsorption- The process by which an ultra thin layer of one substance forms on the surface of another substance.
Alkaline- A substance having a pH greater than 7. See base, or basic. See pH.
Amalgam- An alloy of mercury with another metal or metals.
Amalgamation- The process by which mercury is alloyed with some other metal to produce an amalgam. It was used at one time for the extraction of gold and silver from pulverized ores, but has been superseded by the cyanide process.
Analog- Displaying a readout by a pointer or hands on a dial rather than by numerical digits.
Analysis- The ascertainment of the kind or amount of one or more of the constituents of materials.
Anhydrous- Dry, all water removed, especially the water of crystallization.
Aqua Fortis- Nitric Acid, HNO_3.
Aqua Regia- A mixture of nitric and hydrochloric acids used to dissolve precious metals.
Arsenic- A grayish white element having a metallic luster, vaporizing when heated, and forming poisonous compounds Symbol: As.
Ash- The powdery residue of matter that remains after burning. The process of burning a material to create ash.
Aspirate- To draw in by suction, as used with Atomic Absorption Spectrophotometer.
Assay Ton- A specific weight related to the number of grams in a short ton. An assay ton is 29.1667 grams.
Assay- To analyze (an ore, alloy, etc.) to determine the content of gold, silver, or other metal.
Attrit- The process of wearing loose particles from a material, such as activated charcoal, by gently agitating particles against each other in water.
Bag House- A structure containing filter media to remove contaminants from air, usually dust.
Baking Soda- Sodium Bicarbonate, $NaHCO_3$.
Balance- An instrument for determining weight.
Base metal- Any metal other than a precious or noble metal, such as copper, lead, zinc, or tin.
Basic- A substance having a pH greater than 7. See alkaline. See pH.
Bead- A small ball, or bead, of precious or noble metals remaining in a cupel after cupellation. Part of the fire assay process.
Black Sands- Magnetite or Hematite. Forms of iron or iron compounds recovered as impurities in the placer mining process.

Bond Work Index- This procedure is used to determine power consumption in crushing and grinding to the feed and product size distribution.
Bone Ash- A white ash obtained by roasting, or calcining bones.
Borax Glass- An important flux ingredient made by calcining hydrated sodium borate, $Na_2B_4O_7 \cdot 10H_2O$.
Borax- A white, water-soluble powder or crystals, hydrated sodium borate, that is calcined to remove water and create borax glass. See borax glass.
Borosilicate Glass- See slag.
Bullion- Relatively pure noble metals, considered in mass rather than in value. Usually in bars or ingots.
Bureaucrat- An official who works by fixed routine without exercising intelligent judgment.
Calcine- to convert into calx by heating or burning. See calx.
Calibration- To set or check the graduation of a quantitative measuring instrument.
Calx- The oxide or ashy substance that remains after metals, minerals, etc., have been thoroughly burned.
Cap- To cover, or top off with a layer of flux, borax glass, or other dry reagent.
Carbon- A nonmetallic element found combined with other elements in all organic matter and in a pure state as diamond and graphite Symbol: C. See also activated charcoal. A reducing agent. Any source of carbon, such as flour or sugar can be used as a reducing agent. **Carbonate-** Used to refer to specific types of ores, such as calcium carbonate, or calcite. Carbonates usually have a "CO_3" suffix.
Carbon Dioxide- A colorless, odorless, incombustible gas, CO_2, that is a by- product of smelting carbonate ores.
Carat- A unit of weight in gemstones. Not used for noble metals.
Carcinogen- Any substance or agent that tends to produce a cancer.
Cast- To form (an object) by pouring metal into a mold and letting it harden.
Celsius- Pertaining to or noting a temperature scale in which 0° represents the ice point (freezing) and 100° the steam point (boiling).
Cementation- A chemical process where an inexpensive metal, usually iron or aluminum, is used to cause a chemical reaction that will produce or precipitate a noble metal out of a solution that contains noble metal.
Chemical- a substance produced by, used in, or concerned with chemistry or chemicals.
Chloridize- To convert into chloride; applied to the roasting of silver ores with salt, preparatory to amalgamation.
Collar-The term applied to the timbering or concrete around the mouth or top of a shaft, or the beginning point of a shaft or drill hole, the surface.
Collector- A person or thing that collects. In a smelt, a large enough ratio of metal, that when molten, will collect other metals that are present.
Compound- A substance that is composed of two or more parts, elements, or ingredients.
Concentrate- to separate (metal or ore) from rock, sand, etc., so as to improve the quality of the valuable portion.
Condiment- Something used to flavor food, such as mustard, ketchup, salt, or spices.

Cons- Slang term for concentrate, see concentrate. The end result of the concentrating process.
Contaminate- To inadvertently make impure or unsuitable by contact or mixture with something unclean or bad. To pollute or taint.
Copper- A malleable ductile metallic element having a characteristic reddish brown color. Used in large quantities as an electrical conductor and in the manufacture of alloys, as brass and bronze Symbol: Cu.
Corrosive- Having the quality of corroding or eating away; erosive, such as acid vapors, or solutions.
Crucible Wash- A dry reagent that is lighter than other flux constituents when molten, and therefore is the last portion of the molten material to leave the crucible when poured.
Crucible- A container of refractory material employed for heating substances to high temperatures.
Crystalline- Of or like crystal; clear; transparent.
Cubic Feet Per Minute- A term used to describe the volume of air that is being moved in one minute. CFM.
Cupel- A small, cup-like, porous container, with a hemispherical depression to focus heat in the center. Usually made of bone ash or magnesite, and used in assaying, for collecting gold and silver from lead. The bone ash or magnesite absorb about 90% of the lead, the remainder is vaporized as lead oxide.
Cupellation- To heat or refine in a cupel. The process of removing lead from noble metals in a fire assay.
Cyanide- A salt of hydrocyanic acid, as potassium cyanide, KCN, or sodium cyanide, NaCN. To treat with a cyanide, as an ore, in order to extract gold.
Desulfurizing Agent- A dry reagent used in the fire assay or smelt to remove sulfur. Soda ash is a common reagent used for this purpose.
Dewater- To remove moisture from a slurry by various means, such as a thickener, belt or drum filter, or to remove water from a mine; an expression used in the industry in place of the more technically correct word, unwater.
Digital- Displaying a readout in numerical digits rather than by a pointer or hands on a dial.
Discharge Permit- A permit from a governmental agency allowing the discharge of a specified amount of a toxic or polluting compound from an industrial facility into the environment.
Dore'- An alloy containing gold.
Electrolytic cell- A container with an anode and cathode that a precious metal bearing solution is passed through. Low voltage is passed through the cell to remove the precious metal from the solution.
Elements- One of a class of substances that cannot be separated into simpler substances by chemical means.
Endothermic- Noting or pertaining to a chemical change that is accompanied by an absorption of heat.
Exothermic- Noting or pertaining to a chemical change that is accompanied by a liberation of heat.
Fahrenheit- Noting, pertaining to, or measured according to a temperature scale in which 32° represents the (freezing) ice point and 212° the (boiling) steam point Symbol: F.
Fineness- The proportion of pure precious metal in an alloy, often expressed in parts per thousand.

Fire Assay- An analytical process utilizing heat and dry reagents to quantitatively determine the amount precious metals in an ore. Considered to have a detection limit of .001 OPT.
Firebrick- A brick made of fire clay.
Fire Polish- To repeatedly smelt with an oxidizing flux to increase the fineness of the precious metal content, usually gold.
Flammable- Easily set on fire; combustible.
Flocculent-Particulate in a solution coalescing and adhering in flocks. A cloudlike mass of precipitate in a solution.
Flour- The finely ground meal of grain, especially wheat. Used as a source of carbon, and as a reducing agent in the fire assay.
Flue dust- Dust accumulating in a flue, or ventilation system. May contain very high precious metal values.
Fluorite- A mineral, calcium fluoride, CaF_2, occurring in crystals and in masses: the chief source of fluorine. Also called Fluorspar.
Fluorspar- Calcium fluoride. See Fluorite.
Flux- A substance used to refine metals by combining with impurities to form a molten mixture that can be readily removed. Usually made from dry reagents.
Forceps- An instrument, as in pincers or tongs, for seizing and holding objects firmly, as in surgical operations. Giant tweezers.
Frother- A substance used in a flotation process to make air bubbles sufficiently permanent, principally by reducing surface tension.
Fumes- Any smoke like or vaporous exhalation from matter or substances, especially of an odorous or harmful nature.
Furnace- An apparatus in which heat may be generated, as for heating houses, smelting ores, or producing steam.
Fusion- The act or process of fusing or the state of being fused. To combine or blend by melting together; melt.
Galena- A common heavy mineral, lead sulfide, PbS, occurring in lead-gray crystals, usually cubes, and cleavable masses. The principal ore of lead.
Gangue- the worthless rock or vein matter in which valuable metals or minerals occur.
German Silver- Any of various alloys of copper, zinc, and nickel, usually white and used for utensils and drawing instruments; nickel silver.
Gold- A precious yellow metallic element, highly malleable and ductile, and not subject to oxidation or corrosion Symbol: Au. One of the noble metals.
Grade- A term used in the mining industry to denote values contained in an ore, or other product. There can be high grade, or low grade. To "high grade" can mean to steal.
Graphite- A soft native carbon occurring in black to dark gray foliated masses: used for pencil leads, as a lubricant, as a moderator in nuclear reactors, and for making crucibles and other refractories; also known as plumbago.
Gravimetric- A method of mechanical separation, usually with water, by specific gravity, or referring to specific gravity. Normally, concentrating tables, jigs, centrifuges, sluice boxes, or other devices are used for gravimetric separations.

Gyratory- A widely used form of rock breaker or crusher, in which an inner cone rotates eccentrically in a larger outer hollow cone.
Hallmark- An official mark or stamp indicating a standard of purity, used in marking gold and silver articles.
Head Assay- The assay of head ore, or the assay of the original material before any processing or treatment that may change the characteristics of the original material.
Heat Sink- An environment or medium that absorbs excess heat.
Homogenous- Well mixed or blended. Representative of the whole.
Hydrate- Any of a class of compounds containing chemically combined water.
Hydrofuge- A device that uses centrifugal force, in the presence of water to separate classified materials by specific gravity.
Hygroscopic- Absorbing or attracting moisture from the air.
IRS- The Internal Revenue Service. And they have no sense of humor.
Inert- Having little or no ability to react.
Ingot- A mass of metal cast in a convenient form for shaping, remelting, or refining.
Iodized- To treat, impregnate, or affect with iodine or an iodide.
Inorganic- Noting or pertaining to chemical compounds that are not hydrocarbons or their derivatives.
Inquart- Times four. To add a measured amount. Such as adding four times as much silver to a measured amount of gold to allow complete parting, or chemical separation.
Insulator- A material that absorbs or deflects heat.
Iron- A ductile, malleable, silver-white metallic element, used in it's impure carbon-containing forms for making tools, implements, or machinery. Symbol: Fe. A primary ingredient in steel.
Karat- A unit for measuring the fineness of gold, pure gold being 24 karats fine.
Kiln- A furnace or oven for burning, baking, or drying something, especially one for firing pottery, calcining limestone, or baking bricks.
Kilo- A metric unit of mass, one thousand grams. A Kilogram. Approximately 2.2pounds.
Lead Acetate- a white, crystalline, water-soluble, poisonous solid, $Pb(C_2H_3O_2)$ 2x $3H_2O$. Toxic by ingestion.
Lead- A heavy, comparatively soft, malleable, bluish-gray metal, sometimes found in its natural state but usually combined as a sulfide, as in galena. Symbol: Pb.
Lime- A white or grayish white, odorless, lumpy, very slightly water-soluble solid, CaO, used chiefly in mortars, plasters, and cements, in bleaching powder, and in the manufacture of steel, paper, glass, and various chemicals of calcium.
Limestone- A sedimentary rock consisting predominantly of calcium carbonate, varieties of which are formed from the skeletons of marine microorganisms and coral: used as a building stone and in the manufacture of lime.
Liquefy- To make or become liquid, with the use of high temperature.
Litharge- A yellowish or reddish poisonous solid, PbO, used chiefly in the manufacture of storage batteries, pottery, enamels, and inks. A very important ingredient in fire assay flux.
Manganese Dioxide- A hard, brittle, grayish white, metallic element, an oxide of which, MnO_2, is a valuable oxidizing agent: used chiefly as an alloying agent in strengthening steel. The natural ore of manganese is pyrolusite.

MSHA- (Mine Safety and Health Administration)
Malignant- As in tumor, characterized by uncontrolled growth; cancerous, invasive, or metastatic.
Material Data Safety Sheet- A document produced by a chemical manufacturing company to advise consumers of the hazardous properties of that chemical. Also known as an MSDS.
Mechanical error- An error resulting from mechanical handling, such as a dusting loss when finely divided materials are handled roughly, or poorly.
Mercury- A heavy, silver-white, toxic metallic element, liquid at room temperature, used in barometers, thermometers, pesticides, pharmaceuticals, mirror surfaces, and as a laboratory catalyst; Quicksilver. Symbol: Hg.
Mesh- An arrangement of interlocking metal links or wires with evenly spaced, uniform small openings between, as used in jewelry, sieves, etc.
Metallic- Of, pertaining to, or consisting of metal. Being in the free or uncombined state, such as metallic iron.
Mill- A mineral treatment plant in which crushing, wet grinding, and further treatment of ore is conducted.
Mine- An opening or excavation in the ground for the purpose of extracting minerals; a pit or excavation from which ores or other mineral substances are taken by digging; an opening in the ground made for the purpose of taking out minerals, or a work for the excavation of minerals by means of pits, shafts, levels, tunnels, etc., as opposed to a quarry, where the whole excavation is open.
Mine Safety and Health & Training- MSHA. The Federal Metal and Nonmetallic Mine Training, Safety & Health Standards as defined in 30 CFR, 46-48, 56-58, and 62.
Mining- The science, technique, and business of mineral discovery and exploitation. Strictly, the word connotes underground work directed to severance and treatment of ore or associated rock. Practically, it includes opencast work, quarrying, alluvial dredging, and combined operations, including surface and underground attack and ore treatment.
Mint- A place where coins, paper currency, special medals, etc., are produced under government authority.
Mold (Pouring)- A hollow form for giving a particular shape to something in a molten or plastic state.
Molten- Liquefied by heat; being in a state of fusion.
NAFTA- North American Free Trade Agreement.
Nepotism- Favoritism (as in appointment to a job) based on kinship.
Neutral- Exhibiting neither acid nor alkaline qualities, having a neutral pH.
Neutralize- To make (a solution) chemically neutral. To change pH.
NIMBY- An acronym meaning **N**ot **I**n **M**y **B**ack **Y**ard.
Niter- Potassium Nitrate or Sodium Nitrate. A strong oxidizer and important flux ingredient.
Nitric acid- A colorless or yellowish, fuming, suffocating, water-soluble toxic liquid, HNO_3, used chiefly in the manufacture of explosives and fertilizers.
Nomex- The trade name for a brand of fire proof garments.
OSHA- (Occupational Safety and Health Administration)
Opaque- Not allowing light to pass through.

Ore- A metal-bearing mineral or rock, or a native metal, that can be mined at a profit. A mineral or natural product serving as a source of some nonmetallic substance, as sulfur.
Organic- Noting or pertaining to a class of chemical compounds that formerly comprised only those existing in or derived from plants or animals, but that now includes all other compounds of carbon.
Osmium- A hard, heavy, metallic element, densest of the known elements, able to form octavalent compounds: used chiefly as a catalyst, in alloys, and in the manufacture of electric-light filaments. Symbol: Os. A Platinum Group element.
Ounces Per Ton- OPT. A term indicating the Troy ounces of noble metals in a short ton (2000 pounds) of ore.
Oxidation- The process of adding oxygen to a chemical reaction, usually by the addition of an oxidizer, or oxidizing agent.
Oxide- A compound in which oxygen is bonded to one or more electropositive atoms. A term used to denote a non-complex ore.
Oxidizing agent- A chemical compound that gives oxygen to a chemical reaction.
Part- To separate. Usually refers to the separation of gold and silver by wet chemical means.
Personal Protective Equipment- Personal safety equipment such as respirators, steel toe shoes, lab coats, etc. It is considered the individual's responsibility to use and maintain this equipment, which is normally provided at the workplace.
pH- The symbol for the logarithm of the reciprocal of hydrogen ion concentration in gram atoms per liter, used to describe the acidity or alkalinity of a chemical solution on a scale of 0 (more acidic) to 14 (more alkaline, or basic).
Pin Tube- A piece of glass tubing that has been evacuated of air, designed to sample molten metal. The tube usually has a blister at the end to be immersed in the molten metal. The blister melts at a lower temperature than the tube, allowing the vacuum to pull several inches of molten metal into the glass tube. After cooling, the glass is broken away from the metal, or "pin", which is considered a representative sample of the melt.
Placer- A natural concentration of heavy metal particles, as gold or platinum, in sand or gravel deposited by rivers or glaciers.
Placer Gold- Gold recovered from a placer mining operation.
Platinum Group Elements- Platinum, Palladium, Rhodium, Osmium, Ruthenium and Iridium.
Pollutant- Any substance, as a chemical or waste product, that renders the air, water, or other natural resource harmful or generally unusable.
Portal- The surface entrance to a drift, tunnel, adit, or entry. Or the log, concrete, timber, or masonry arch or retaining wall erected at the opening of a drift, tunnel, or adit.
Potassium Nitrate- A crystalline compound, KNO_3, produced by nitrification in soil, and used in gunpowder, fertilizers, and preservatives; saltpeter; niter. A strong oxidizer in fluxes.
Precious Metal(s)- A metal of the gold, silver, or platinum group.
Precipitate- To separate (a substance) in solid form from a solution, as by means of a reagent.
Preponderance- The fact or quality of being preponderant; superiority in weight, power, numbers, etc. More of.
Proprietary- Manufactured and sold only by the owner of the patent, trademark or process. Closely held.

Protective Alkalinity- A basic, or alkaline pH range used in cyanide operations, typically 9.0 to 10.5. Cyanide will not evolve hydrogen cyanide gas as long as this pH range is maintained, usually with sodium hydroxide.
Pyrolusite- A grayish black mineral, manganese dioxide, MnO_2, the principal ore of manganese.
Qualitative- Pertaining to or concerned with quality. Qualitative analysis normally will indicate what elements are present, but not how much of the element is present.
Quantitative- Being measured by quantity. Quantitative analysis will indicate precisely how much of a single element is present.
Quench- To cool suddenly by plunging into a liquid, as in tempering steel by immersion in water.
Readability- Pertaining to the accuracy and weight range of an instrument such as a balance, scale, or other weighing device.
Reagent- A substance that, because of the reactions it causes, is used in analysis and synthesis.
Reducing Agent- A substance that causes another substance to undergo reduction and that is oxidized in the process. A source of carbon, such as flour.
Refine- To bring to a pure state; free or separate from impurities or other extraneous substances.
Refinery- An establishment for refining something, as metal, sugar, or petroleum.
Refining- The process of bringing to a pure state. The process of separating impurities.
Refractory- A material that retains its shape and composition even when heated to extreme temperatures.
Refractory Ore- An ore that is difficult to fuse, reduce, or work.
Representative- A typical example or specimen. A small portion that represents the whole.
Residual- Pertaining to or constituting a residue or remainder; remaining; leftover.
Respirator- A filtering device worn over the nose and mouth to prevent inhalation of noxious substances.
Retort- A vessel, commonly a metal chamber with a long neck bent downward, used for distilling or decomposing substances by heat. A device for separating gold and mercury (an amalgam) from one another.
Roasting Dish- A refractory container, usually round, used for roasting (oxidizing) a mineral sample.
Salt- A crystalline compound, sodium chloride, NaCl, occurring chiefly as a mineral or a constituent of seawater, and used for seasoning food and as a preservative.
Saltpeter- Naturally occurring potassium nitrate, used in making fireworks, gunpowder, etc.; niter.
Salting- The fraudulent adulteration of a sample, for example, adding a small amount of gold to a sample to make it appear that the gold content of the rock is much higher than it actually is. Salting may be accidental, caused by the fortuitous segregation of rich mineral during sampling. Sampling methods are conducted to reduce chance segregation to a minimum.
Sand- The more or less fine debris of rocks, consisting of small, loose grains, often of quartz.
Self Contained Breathing Apparatus- SCBA- A breathing device that supplies compressed air, as opposed to a respirator, which filters air.
Shaft- A vertical or inclined opening of uniform and limited cross section made for finding or mining ore, or ventilating underground workings.
Shotted- Containing small, round metallic particles.

Silica- The dioxide form of silicon, SiO_2, occurring as quartz sand, flint, and agate: used chiefly in the manufacture of glass, water glass, ceramics, and abrasives. Also called silicon dioxide.
Silicosis- A disease of the lungs caused by the inhaling of siliceous particles, as by stone cutters or miners.
Silmanite- A refractory compound used to manufacture vessels for use at high temperatures.
Silver Chloride- A white powder, AgCl, that darkens on exposure to light: used chiefly in photographic emulsions and in antiseptic silver preparations.
Silver- A white, ductile metallic element, used for making mirrors, coins, ornaments, table utensils, photographic chemicals, and conductors. Symbol: Ag.
Slag Pot- A tapered, heavy metal container used to contain smelted metals. The taper allows for separation of the metal from the slag.
Slag- The more or less completely fused and vitrified matter separated during the reduction of a metal from its ore. Borosilicate glass containing the impurities from a smelt.
Slake- To cause disintegration by treatment with water.
Smelting- The process of fusing or melting in order to separate metal contained. To obtain or refine (metal) in this way.
Soda Ash- Sodium Carbonate, Na_2CO_3.
Sodium Bicarbonate- A white water-soluble powder, $NaHCO_3$, used chiefly as an antacid, a fire extinguisher, and a leavening agent in baking. Also called bicarbonate of soda, baking soda. Evolves large amounts of gas at high temperature, not considered a useful flux ingredient.
Sodium Carbonate- Also called soda ash. An anhydrous, grayish white, odorless, water-soluble powder, Na_2CO_3, used in the manufacture of glass, ceramics, soaps, paper, petroleum products, sodium salts, as a cleanser, for bleaching, and in water treatment. A valuable flux ingredient.
Sodium Chloride- See salt. A crystalline compound, sodium chloride, NaCl, occurring chiefly as a mineral or a constituent of seawater, and used for seasoning food and as a preservative.
Spall- To violently break or split off in chips or bits.
Spot- The daily fixed price of gold and other commodities.
Sprout- To spontaneously erupt. A phenomena of molten silver at high temperature and cooling.
Stack Permit- See discharge permit.
Static Pressure- The resistance to the flow of air through duct work or piping that must be overcome by a blower.
Sterling Silver- Silver having a fineness of 0.925, now used in the manufacture of table utensils, jewelry, etc.
Stoney- Resembling or suggesting stone.
Sulfide- A compound of sulfur with a more electropositive element or, less often, group. Such as Iron Pyrite.
Sulfur Dioxide- A colorless, nonflammable, water-soluble, suffocating gas, SO_2, formed when sulfur burns: used chiefly in the manufacture of chemicals such as sulfuric acid, in preserving fruits and vegetables, and in bleaching, disinfecting, and fumigating.
Sulfuric Acid- A clear, colorless to brownish, dense, oily, corrosive, water miscible liquid, H_2SO_4, used chiefly in the manufacture of fertilizers, chemicals, explosives, and dyestuffs and in petroleum refining. Also called Oil of Vitriol.

Surfactant-A surface active agent, a substance that affects the properties of the surface of a liquid or solid by concentrating in the surface layer.

Suspended- To keep from falling or sinking, as if by hanging. To suspend particles in a liquid.

Thallium- A soft, malleable, bluish white metallic element. Used in the manufacture of alloys and, in the form of its salts, in rodenticides. Extremely toxic in some forms. Symbol: Tl.

Thermal Shock- Stress to a refractory container, such as a crucible, caused by heating to extreme temperatures and cooling.

Tilting Furnace- A large furnace that tilts on an axis as it is elevated to the pour position by mechanical means.

Touchstone- A black stone once used to test gold and silver by rubbing them on it. Used to refer to a streak (color) test.

Toxic- Acting as or having the effect of a poison. Harmful to the human body.

Toxicity- The quality, relative degree, or specific degree of being toxic or poisonous.

Translucent- Permitting light to pass through but diffusing it so that objects on the opposite side are not clearly visible. Frosted window glass is translucent.

Trommel-A revolving cylindrical screen used in size classification of coarsely crushed ore, coal, gravel, and crushed stone. The material to be screened is delivered inside the trommel at one end. The fine material drops through the holes; the coarse material is delivered at the other end.

Troy Weight- A system of weights in use for precious metals and gems, in which a pound equals 12 ounces (0.373 kg) and an ounce equals 20 pennyweights or 480 grains (31.1035 grams).

Uniodized- Does not have iodine added.

Unslaked- Not treated with water, referring to lime.

Upgrade- To improve or enhance the quality or value of a precious metal bearing material by chemical or gravimetric means.

Vapor- A substance in gaseous form that is below its critical temperature. Usually toxic if produced a high temperatures.

Ventilation- Facilities or equipment for providing ventilation. The process of moving air through an enclosed area to remove vapors, fumes, or dust.

Viscous- Of a glutinous nature or consistency; sticky; thick; stringy; adhesive.

Washing Soda- See sodium carbonate.

Winze- A subsidiary shaft that starts underground. It is usually a connection between two levels.

Zinc- A ductile, bluish white metallic element: used in making galvanized iron. brass, and other alloys, and as an element in voltaic cells. Symbol: Zn.

Appendix A

How to Read A Material Safety Data Sheet (MSDS)

On the next several pages is our sample MSDS. Take the time to read the document, and understand the information that is provided for you in the document. This information can save you a lot of pain, perhaps even death, so take it seriously.

Some of the terminology can be a bit rough, so here are some terms you will see in the document, and what they mean:

Boiling Point: The temperature at which a liquid starts to boil, or changes to vapor.

Explosive Limits: The minimum and maximum concentration above and below which a substance will not explode.

Flammable Limits: The minimum and maximum concentrations above and below which a chemical won't catch fire.

Hazardous Polymerization: Some chemicals can have a reaction with themselves that will result in an explosion. Your MSDS should explain to you how to keep this from happening.

Melting Point: The temperature at which a solid changes into a liquid.

Permissible Exposure Limits (PEL'S): The maximum concentration of an air contaminant that workers may be exposed to without suffering adverse effects to their health. The limits are established and enforced by OSHA. Personal protective equipment will generally be required if concentrations are above the established PEL.

Specific Gravity: The weight of the material compared to the weight of an equal volume of water. Water has a specific gravity of one. If the chemical has a higher specific gravity, it will sink in water, if the specific gravity I s lower than water, it will float on water.

Stability: The ability of a substance, or chemical, to remain unchanged. If the chemical is listed as unstable, information should be provided regarding what conditions should be avoided to prevent hazards, such as heat, pressure, water, contact with organic materials, etc.

Threshold Limit Values (TLV'S): Similar to PEL's, except that these limits are recommended limits set by the American Conference of Governmental Industrial Hygienists, known as ACGIH.

Vapor Density: This number will indicate the density of the chemical's vapor, with air being one. If the number is higher than one, the vapor will sink (heavier than air), and if the number is lower than one, the vapor will rise (lighter than air).

Vapor Pressure: This number indicates how easily the chemical evaporates, or releases vapor. The higher the number, the faster it will evaporate.

Time Weighted Average (TWA): The permissible amount of chemical one may be exposed to, usually expressed in milligrams per cubic meter per an eight hour period.

And here we go....

Your Favorite Chemical Company, 123 Street, Anytown, USA

Material Safety Data Sheet

Chemtrec # (800) 424-9300---National Response Center # (800) 424-8802

Effective Date: 3-25-96 **CALCIUM OXIDE** Issued 3-25-96

Section One-Product Identification

Product Name: . . . . Calcium Oxide
Common Synonyms:. . . . . Lime, Calx, Quicklime
Calcium Monoxide, Burnt Lime
Chemical Family: . . . . Calcium Compounds
Formula: . . . . CaO
Formula Weight: 56.08
CAS Number: 1305-78-8
NIOSH/Rtecs Number:. EW3100000
Product Use: . Laboratory Reagent

Precautionary Labeling

Safety Data System:
Health . . . . 1 Slight
Flammability. . . . . 0 None
Reactivity . . . . 1 Slight
Contact . . . . 2 Moderate
Laboratory Protective Equipment: . . . . Goggles, Lab Coat

US Precautionary Labeling:

W A R N I N G

Causes irritation. Harmful if swallowed.
Avoid Contact with eyes, skin, clothing. Keep in tightly closed container.
Wash thoroughly after handling.

International Labeling:

Irritating to eyes, respiratory system and skin. Irritating to eyes and skin. Avoid contact with eyes. After contact with skin, wash immediately with plenty of water. Keep container tightly closed.

Storage color code: Orange (General Storage)

Section II - Components

Component:. .	CAS NO.	Weight %	OSHA/PEL	ACGIH/TLV
Calcium Oxide	1305-78-8	99-100	5 Mg/M^3	N/E

Section III - Physical Data

Boiling Point: 2850 C (5162 F) Vapor Pressure: (MM Hg) N/A (at 760 MM Hg)
Melting Point: 2572 C (4661 F) Vapor Density: (Air=1) 1.9
Specific Gravity: 3.34 Evaporation Rate: N/A
Solubility (H_2O): Negligible (<0.1% % Volatiles By Volume: 0
Odor Threshold (PPM): N/A Physical State: Solid
Coefficient Water/Oil Distribution: N/A
Appearance & Odor: White to Gray Solid. Odorless.

Section IV - Fire And Explosion Hazard Data

Flash Point (Closed Cup): N/A NFPA 704M Rating: 1-0-1
Auto ignition Temperature: N/A
Flammable Limits: Upper - N/A Lower - N/A
Fire Extinguishing Media: Use Extinguishing Media Appropriate for Surrounding Fire.
Special Fire Fighting Procedures: None Identified.
Unusual Fire & Explosion Hazards: Contact With Moisture or Water May Generate Sufficient Heat to Ignite Combustible Materials.
Toxic Gases Produced: Non Identified.
Explosion Data - Sensitivity to Mechanical Impact: None Identified.
Explosion Data - Sensitivity to Static Discharge: None Identified.

Section V - Health Hazard Data

Threshold Limit Value (TLV/TWA): 2 Mg/M^3
Short Term Exposure Limit (STEL): Not Established
Permissible Exposure Limit: 5 Mg/M^3
Toxicity of Components: No Information Available.
Carcinogenicity: NTP: No IARC: No Z List: No OSHA Reg: No
Carcinogenicity: None Identified.
Reproductive Effects: None Identified.
Effects of Overexposure:

Inhalation: Tightness and pain in chest, coughing, difficult breathing.
Skin Contact: Severe irritation or burns.
Eye Contact: Severe irritation or burns.
Skin Absorption: None identified.
Ingestion: Irritation and burns to mouth and stomach.

Chronic Effects: None identified.

Target Organs: Respiratory System, Lungs, Kidneys, Prostate, Blood.
Medical Conditions Generally Aggravated By Exposure: None Identified.
Primary Routes of Entry: Inhalation, Ingestion, Skin Contact, Eye Contact.
Emergency and First Aid Procedures:
Ingestion: Call a physician. If swallowed, do not induce vomiting. If conscious, give large amounts of water. Follow with diluted vinegar, fruit juice, or whites of egg beaten with water.
Inhalation: If inhaled, remove to fresh air. If not breathing, give artificial respiration. If breathing is difficult, give oxygen.
Skin Contact: In case of contact, flush skin with water.
Eye contact: In case of eye contact, immediately flush with plenty of water for at least fifteen minutes.

SARA/Title III Hazard Categories and Lists:

Acute: Yes Chronic: Yes Flammability: No Pressure: No Reactivity: No
Extremely Hazardous Substance: No CERCLA Hazardous Substance: No
SARA 313 Toxic Chemicals: No TSCA Inventory: Yes

Section VI - Reactivity Data

Stability: Stable
Hazardous Polymerization: Will Not Occur.
Conditions To Avoid: Moisture, Air.
Incompatibles: Water, Fluorine, Strong Acids.
Decomposition Products: None Identified.

Section VII - Spill & Disposal Procedures

Steps to be taken in the event of a spill or discharge: Wear self-contained breathing apparatus, and full protective clothing. With clean shovel, carefully place material into a clean, dry container and cover; remove from area. Flush spill area with water.

Disposal Procedure: Dispose in accordance with all applicable Federal, State, and Local environmental regulations.

Section VIII - Industrial Protective Equipment

Ventilation: Use general or local exhaust ventilation to meet TLV requirements.

Respiratory Protection: Respiratory protection required if airborne concentration exceeds TLV. At concentrations up to 11 PPM, a dust/mist respirator is recommended. Above this level, a self contained breathing apparatus is advised.
Eye/Skin Protection: Safety goggle, uniform, and rubber gloves are recommended.

Section IX - Storage & Handling Precautions

SAF-T-Data Storage Color Code: Orange (General Storage)

Storage Requirements: Keep tightly closed. Suitable for any general chemical storage area. Store in a dry area.

Section X - Transportation Data & Additional Information

Domestic (DOT):

Proper Shipping Name: Calcium Oxide (Air Only)
Hazard Class: 8 Packaging Group: III
UN/NA: UN1910 Labels: Corrosive
Regulatory References: 49 CFR 172.101

International (IMO)
Proper Shipping Name: Chemicals, NOS (Non-Regulated) Calcium Oxide (Material Hazardous Only in Bulk) Marine Pollutants: No
Air (ICAO)
Proper Shipping Name: Calcium Oxide Hazard Class: 8 UN: UN1910
Labels: Corrosive Packaging Group: III
Regulatory References: 49 CFR 172.101; 173.6 ; Part 175; ICAO/IATA–

We believe the transportation data and references contained herein to be factual and the opinion of qualified experts. The data is meant as a guide to the overall classification of the product and is not package size specific, nor should it be taken as a warranty or representation for which the company assumes legal responsibility.

The information is offered solely for your consideration, investigation, and verification. Any use of the information must be determined by the user to be in accordance with applicable Federal, State, and Local laws and regulations. See shipper requirements 49CFR 172.3 and employee training, 49CFR 173.1.

-End-

The Anatomy of a Tilting Furnace

-One-

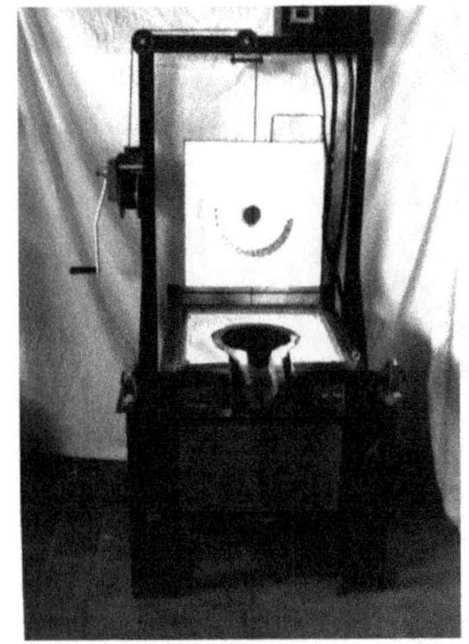

-Two-

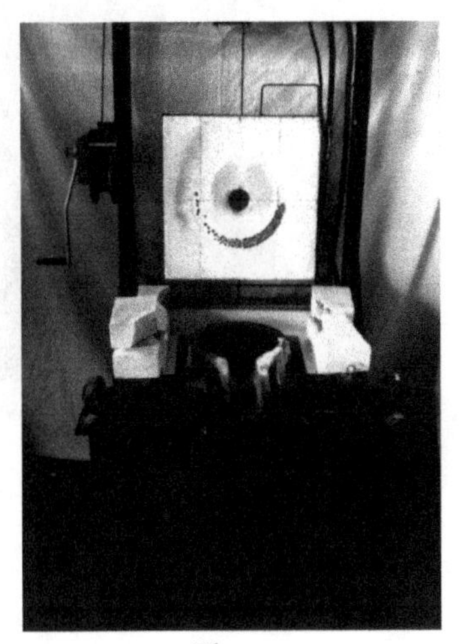

-Three-

-Four-

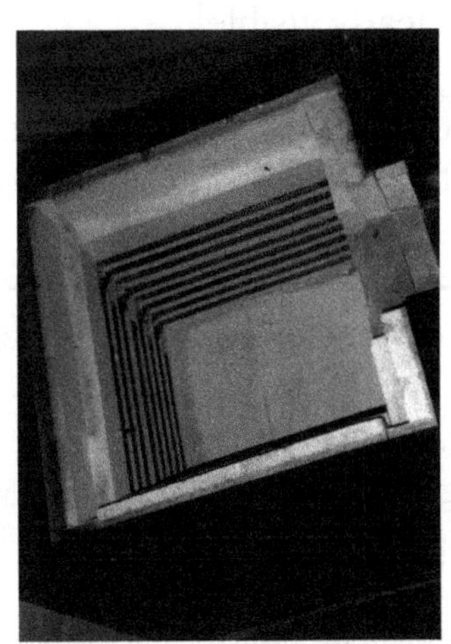

-Five-

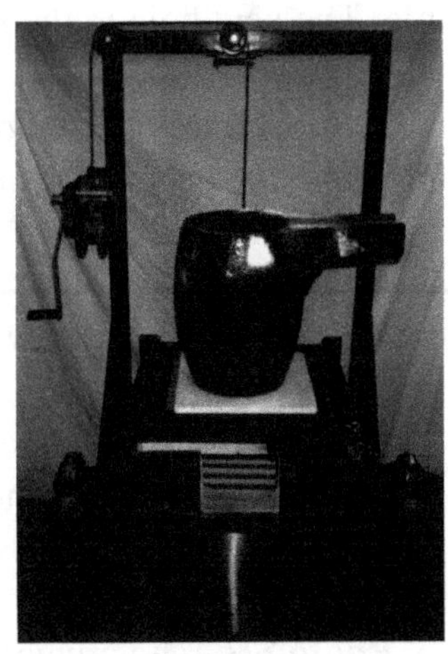

-Six-

-Seven-

-Eight-

Photo One-This is a photo of a Vcella TL-60 Tilting Furnace. The controller is at the upper right. The handle on the left is for the winch assembly that raises the box containing the crucible. The nose of the silicon carbide crucible is visible at center front of the box.

Photo Two- This photo is with the top open. The retaining bricks are visible around the crucible. The vent hole in the top is clearly visible.

Photo Three- This photo shows the retaining bricks removed from around the top of the crucible. The crescent shape on the top is where the glaze on the crucible was in contact with the brick in the top.

Photo Four- The view into the furnace box with the top brick removed.

Photo Five-The furnace box, empty. The notch for the crucible spout is to the right, the grooves in the sides hold the heating elements, and the thermocouple is visible on the upper left rear of the furnace box.

Photo Six- The crucible, with the attached base are sitting on the box with the lid closed. Heating elements are visible through the notch for the crucible spout.

Photo Seven- The tilting furnace in the "pour" position. Note how the top front of the box is the pivot point. The framework supports the box, and provides a pivot point.

Photo Eight- The tilting furnace in the"pour" position, with the top up to show retaining brick. Actually, a very simple mechanism.

Suppliers

The suppliers listed below carry the supplies, equipment, chemicals and other things you will need. There are, no doubt, other suppliers, possibly in your area. Find the closest supplier you can, at a reasonable price. Freight charges can exceed the cost of the merchandise if you aren't careful in choosing your supplier.

Always get a quote before you buy. You can also have the freight rates checked, and specify the method of shipment. Caveat emptor! Remember that chemicals can not be returned, so make sure you know what you are ordering beforehand.

If you are a Nevada resident, note that freight is considered part of the purchase price, and will be taxed according to the county you live in. See NAC 372.101 for more information.

Neither the author, or publisher of this book endorse any of these suppliers!

Hunter Refractories, Inc.
1095 Spice Island Drive #103
Sparks, NV 89431
Phone (775) 355-8300

Hunter supplies crucibles and cupels to the Nevada mining industry. They carry AP Green assay crucibles, which are very popular. They are worth checking with.

Legend, Inc.
988 Packer Way
Sparks, NV 89431-6441
Phone: (775) 786-3003
Fax: (775) 786-3613
http://www.lmine.com.

Legend has all the assay supplies, hardware, books, tongs, molds, and things you might need. They used to publish a catalog, give them a call. They do have used equipment, and they do sell pre-mixed fluxes, such as the "Chapman" gold smelting flux in this book. Be sure to check freight rates.

DFC Ceramics, Inc.
PO Box 110
Canon City, CO 81212-0110
Phone (719) 275-7525
www.dfcceramics.com.

DFC manufactures crucibles for both assay and smelting purposes. Their fused silica crucibles are excellent, as are their assay crucibles. They also manufacture assay furnaces, molds, and other handy items. Call for a catalog, you won't be disappointed. They do have satellite locations, which could save on the freight.

Anachemia Mining LLC
738 Spice Island Drive
Sparks, NV 89431 - View Map
Phone: (775) 331-2300

Anachemia is a full line chemical supplier to the mining industry. Most large chemical companies do not sell to individuals due to liability. Check with them on this, they also have an excellent catalog and inventory.

Action Mining Services, Inc.
37390 Ruben Lane.
Sandy, OR 97055 United States
(800) 624-1511
(503) 826-9330
www.actionmining.com

Action has all sorts of goodies, most of which are very good. They are important because they sell chemicals in small quantities. They have an excellent catalog, and carry a lot of good books. They are very pleasant to deal with, and have a very knowledgeable staff. They serve the mining community worldwide, and have been in business since 1979. The prices are competitive, and they ship fast. They assay, and do spectrographic analysis.

Vcella Kilns, Inc.
171 Mace St., Unit B
Chula Vista, CA 91911
Phone (619) 427-2550

Talk to Phil Strona. These kilns are made primarily for the ceramics industry. A lot of assayers have bought them, and have had excellent results. Vcella also manufactures tilting furnaces that are excellent. We have seen this equipment in operation for years without burning out a heating element. They are the best buy that we know on the market. The kilns and tilting furnaces will hold 2300°F continuously. Parts are very reasonable. Call for a catalog or price list. On the web, check www.vcella-kilns.com.

David H. Fell & Co., Inc.
6009 Bandini Blvd.
City of Commerce, CA 90040
Phone (213) 722-9992
Fax (213) 722-6567
Phone toll free (800) 822-1996

David Fell & Company has been around since 1973, and will buy your precious metal. They seem to have very reasonable rates, and advertise a two day settlement, which is fast. Call for a brochure, or see their online brochure at www.dhfco.com. Their website also has a refining value estimator. Start here to find a refiner.

Keene Engineering
20201 Bahama St.
Chatsworth, CA 91311
Phone (818) 993-0411 Fax (818) 993-0447

Keene is the original prospector's toy store. They are a full line manufacturer of some of the best equipment available at the best price. If you are a placer miner, and you don't have a Keene catalog handy, well, you just blew it. If you are into hard rock, well, better get a catalog anyway, you'll need it. A great source for books, equipment, dredges, high bankers, sluice boxes and a whole lot of other goodies. Keene has a reputation for quality, and they have been around a long, long time. Good stuff. Call for a free catalog, or go to www.keeneeng.com on the Web.

Miners, Inc.
35 Pollock Rd.
PO Box 1301
Riggins, ID 83549-1301
Toll Free (800) 824-7452
Phone (208) 628-3247 Fax (208) 628-3749

Miner's has been around for 34 years, and has shipped merchandise to 115 countries. Here we go: Sample storage & Identification, Sampling Equipment, Magnifiers, Mapping Scales, Surveying, Hand Tools, Compasses, Stereoscopes, Altimeters, Clinometers, Field Books, Leather Field Equipment, Field Apparel, First Aid, Lighting, Microscopes, Laboratory, Gold Panning, Radiation-Ultraviolet, Books, and Gift Ideas. Oh boy! A very good source for professional geologists, but they welcome the novice as well. Catalogs are free, and will have a stunning mineral specimen on the cover. Miner's has an excellent selection of technical books. On the Web, check www.minerox.com.

International California Mining Journal
The Prospecting and Mining Journal
PO Box 2260
Aptos, CA 95001-2260

This is the premiere publication for prospectors, dredgers, and any person interested in mining. There are excellent articles on all subjects related to mining. There are legal updates on mining and dredging information, lots of good ads on new equipment, used equipment, supplies and other mining related goods and services. Electronic subscriptions are available, as are printed magazines. There is a lot of useful information on their web page at www.icmj.com.

MiningBooks.com
www.miningbooks.com

Mining books has books! If you are looking for a reference book, or an old book that has been reprinted, look here. They also have a lot of current books, as well. Very convenient to use, easy to look for your book of interest. The best of mining related books....Period.

Make Your Own Gold Bars
(& Mining Supplies)
www.makeyourowngoldbars.com
1498 E. Main St. Ste 103-258
Cottage Grove, Or 97424
Office 541-942-9994
These folks have got most everything you will need to smelt. All the equipment, tools, books, and other mining related stuff. Be sure to check the website or give them a call.

There are a lot of advertisements in the mining trade publications. You can pick up an issue at the newsstand, or subscribe.

Always check for local sources first to offset freight costs. The Yellow Pages or the Web are always a good place to start.

Useful Conversions

One Troy Ounce = 31.1035 Grams
480 Grains
20 Pennyweight
1.0971 Avoirdupois Ounces

One Troy Pound = 12 Troy Ounces

One Avoirdupois Ounce =28.3495 Grams
437.500 Grains
18.2292 Pennyweight
0.9115 Troy Ounces

One Avoirdupois Pound=16 Ounces

1% = 10,000 PPM (PPM = Parts Per Million)
1 PPM = 1000 PPB (PPB = Parts Per Billion)
1 PPM = .029166 Troy Ounces Per Ton
One Short Ton = 2000 Pounds
One Short Ton = 29,1666 Troy Ounces
One Metric Ton = 1000 Kilograms = 2204.6 Pounds
One Troy Ounce / Short Ton = 34.2857 Grams / Metric Ton or 34.2857 PPM

N

O

P

Q

R

S

Your comments are welcome! We appreciate your business!

Please feel free to write with your suggestions or comments.

If you would like to see books of a different subject matter published, please send your suggestions or information to the author at the address below.

Bear in mind that this book was written and published to fill a specific niche in the mining industry. Hopefully, it has provided specific, useable information for you, the reader.

Good luck with your project.

Send suggestions or comments to:

Hank Chapman, Jr.
2795 Avenida Grande
Bullhead City, AZ 86442
hchap@suddenlink.net

Notes

CPSIA information can be obtained
at www.ICGtesting.com
Printed in the USA
BVHW051351210819
556415BV00023B/1398